"十三五"普通高等教育本科规划教材
全国本科院校机械类创新型应用人才培养规划教材

数 控 技 术
（双语教学版）

Computer Numerical Control Technology

主　编　吴瑞明
副主编　胡　平　吴蒙华　李雄兵

内 容 简 介

本书用英文系统地介绍了数控技术的发展、数控分类和编程技术、计算机数控装置、数控机床的伺服控制和位置检测、数控机床的机械结构和刀具系统、最新 CNC 技术等知识，附录提供了中文实验指导，以便读者对照参考。

本书可作为数控技术应用、机电一体化、机械制造及其自动化、模具设计与制造等专业的数控技术双语教学用书和专业英语教学用书，也可作为数控技术等相关专业技术人员的英语参考用书。

图书在版编目（CIP）数据

数控技术：双语教学版/吴瑞明主编. —北京：北京大学出版社，2017.3
（全国本科院校机械类创新型应用人才培养规划教材）
ISBN 978-7-301-27920-5

Ⅰ. ①数…　Ⅱ. ①吴…　Ⅲ. ①数控技术—双语教学—高等学校—教材　Ⅳ. ①TP273

中国版本图书馆 CIP 数据核字(2017)第 006379 号

书　　　　名	数控技术（双语教学版）
	Shukong Jishu
著作责任者	吴瑞明　主编
策划编辑	童君鑫
责任编辑	黄红珍
标准书号	ISBN 978-7-301-27920-5
出版发行	北京大学出版社
地　　　　址	北京市海淀区成府路 205 号　100871
网　　　　址	http://www.pup.cn　新浪微博:@北京大学出版社
编辑室邮箱	pup6@pup.cn
著作责任者	zpup@pup.cn
电　　　　话	邮购部 010-62752015　发行部 010-62750672　编辑部 010-62750667
印　刷　者	北京虎彩文化传播有限公司
经　销　者	新华书店
	787 毫米×1092 毫米　16 开本　15.5 印张　358 千字
	2017 年 3 月第 1 版　2023 年 8 月第 4 次印刷
定　　　　价	39.00 元

未经许可，不得以任何方式复制或抄袭本书之部分或全部内容。
版权所有，侵权必究
举报电话：010-62752024　电子信箱：fd@pup.pku.edu.cn
图书如有印装质量问题，请与出版部联系，电话：010-62756370

Preface

Computer Numerical Control (CNC) technology integrates computer technology, automatic control technology, information technology, sense technology, and machining technology. It is the fundament to realize the automation, flexibility and integration of the production in manufacturing industry. CNC technology has become one of the most important technical courses for mechanical engineering specialty.

This book introduces the basic components and the control principle of CNC machine tools, the CNC part programming (manual part programming and automatic part programming), the position measuring devices, the computer numerical control unit, and the development of NC technology and automation of manufacturing. More specifically, this book is intended for the following readers:

(1) Academic readers: This book will provide instructors and students a very informative introduction of CNC applying, various machines, and their uses, along with the necessary tools used in the process.

(2) Teachers looking for CNC teaching material: This book can be picked as a bilingual teaching material to help students understand CNC concepts and manufacturing processes.

(3) Fans: There are a great number of fans interested in the understanding and technical aspects of CNC, but are not exactly sure where to begin, what is absolutely required for the application at hand from both a hardware and software perspective and what is not.

(4) Readers looking for an industry guide: This book is also intended to be used as a guide, showing the reader that there are certain industry standards within the field of CNC that should be adhered to. There are proprietary hardware and software systems for sale and this book advises the readers as to the pitfalls of using components and systems that are nonstandard. Furthermore, the readers are armed with the appropriate questions to ask vendors when trying to determine the best approach to take.

What we recommend you is to use a highlighter to help you denote specific items so that you could find the key to understanding the CNC concepts. Start compiling your own listing of values you are looking for: feed rates, spindle speeds, and cut depths for certain tooling and materials, conventional or climb milling orientations for various material types you encounter, tips and tricks to help you remember various software parameters, etc. It may take you some time to find the optimum cutting parameters for a certain type of material.

The book is edited by Associate Prof. Wu Ruiming (Editor-in-chief, Zhejiang University of Science and Technology), Associate Prof. Hu Ping (Wuhan University, Post doctorate, University of Nebraska-Lincoln), Prof. Wu Menghua (Dalian University), Prof.

Li Xiongbing (Central South University). Thanks for help of Prof. Joseph A. Turner (English Proofreading, University of Nebraska-Lincoln). Thanks to team members, Senior engineer Chen Jiansong (Southeast University), Prof. Wu Jian (Zhejiang University of Science and Technology), Senior experimentalist Hu Weirong (Zhejiang University of Science and Technology), Engineer Ling Wei (Zhejiang University of Science and Technology), Wang Lihang (Postgraduate), Wang Fei (Postgraduate).

This book is supported by the State Scholarship Fund under Grant No. 201207570001. It is completed by the cooperation of Zhejiang University of Science and Technology and University of Nebraska-Lincoln.

Editor
October 2016

前　　言

　　计算机数控技术（简称数控技术）包括计算机技术、自动化控制技术、信息技术、传感技术和制造加工技术。它是实现工业自动化、柔性制造和制造信息化的基础。同时，数控技术也是机械类专业的一门重要的专业课程。

　　本书介绍了数控机床的基本组成和控制原理、数控编程方法（包括手工编程和自动编程）、位置测量装置、计算机控制单元和数控技术的发展及自动化制造系统。本书适用于以下读者：

　　(1) 有学术需要的读者：本书介绍了CNC的应用，各类数控机床及其应用，同时介绍了加工刀具知识。

　　(2) 有教学需要的读者：本书可作为双语教学用书，帮助学生理解CNC的概念和加工工艺，有助于学生阅读英文说明书。

　　(3) 对数控技术有兴趣的读者：很多读者对数控技术感兴趣，但不知道如何着手。本书从硬件结构和数控编程两方面入手，期待为数控技术爱好者回答数控技术是什么和为什么的问题。

　　(4) 想了解数控发展和选型的读者：本书可为期望了解数控领域标准的读者提供指导。市面上销售的数控机床硬件和软件系统各式各样，有标准的，也有非标准的。希望本书对人们了解数控部件和整机选型（英文）及其销售有所帮助。同时，对于选用的设备，本书对销售者提出了技术要求。

　　编者还希望本书能够帮助读者更好地理解数控技术的概念和关键技术。通过自身的经验，更好地选用进给量、主轴转速、不同刀具和材料的切削深度，解决不同材料铣削等工程问题，在数控加工时能够更好地进行切削参数的优化。

　　本书由浙江科技学院吴瑞明副教授担任主编，美国内布拉斯加大学林肯分校博士后兼武汉大学胡平副教授、大连大学吴蒙华教授、中南大学李雄兵教授担任副主编，美国内布拉斯加大学林肯分校 Joseph A. Turner 教授负责英文校稿。参加编写的人员还有东南大学陈建松高级工程师，浙江科技学院吴坚教授、胡伟蓉高级实验师、凌玮工程师。研究生王黎航、王飞完成部分资料的收集和校对。

　　本书出版由国家留学基金（编号：201207570001）资助。编者期待与美国内布拉斯加大学林肯分校的进一步合作。

编　者

2016 年 10 月

Contents

Chapter 1 Introduction of CNC ········ 1

   1.1 History of NC Development ········ 1
   1.2 Concept of NC and CNC ········ 2
   1.3 Classifications of CNC Machines ········ 8
   1.4 CNC Application ········ 14
   Exercises ········ 19

Chapter 2 CNC Part Programming ········ 20

   2.1 Introduction ········ 20
   2.2 The Basis of CNC Part Programming ········ 24
   2.3 Definition of Programming ········ 28
   2.4 Part Programming ········ 32
   2.5 Computer Aided Manufacturing ········ 64
   Exercises ········ 69

Chapter 3 CNC Unit and Control Principle ········ 74

   3.1 Hardware Architecture of a CNC Unit ········ 74
   3.2 CNC System Software ········ 81
   3.3 Interpolation ········ 84
   3.4 Forms of Compensation ········ 86
   3.5 CNC Acceleration/Deceleration Control ········ 91
   3.6 PLC Function ········ 93
   Exercises ········ 95

Chapter 4 Servo System and Position Measuring Device ········ 97

   4.1 Introduction ········ 97
   4.2 Servo Systems ········ 99
   4.3 Requirements and Classifications for Position Measuring Devices ········ 109
   4.4 Position Measuring Devices ········ 111
   Exercises ········ 115

Chapter 5 Mechanical Construction and Tool System of CNC Machines ········ 116

   5.1 CNC Machine Tools ········ 116

5.2　Main Structure of CNC Machine Tools …………………………………………… 123

5.3　CNC Tool System ………………………………………………………………… 131

　　Exercises ………………………………………………………………………………… 143

Chapter 6　Architecture for Modern CNC Technology …………………………… 146

6.1　Open Architecture System for CNC Unit ………………………………………… 146

6.2　STEP-NC System ………………………………………………………………… 148

6.3　Advanced Application of CNC Technology ……………………………………… 150

　　Exercises ………………………………………………………………………………… 163

Appendix Ⅰ　Glossary …………………………………………………………………… 164

Appendix Ⅱ　MAHO 数控系统实验指导 ……………………………………………… 185

Ⅱ.1　MAHO 数控系统 ………………………………………………………………… 185

Ⅱ.2　MAHO GR350C 车削编程 ……………………………………………………… 188

Ⅱ.3　MAHO 600C 铣削编程 ………………………………………………………… 198

Appendix Ⅲ　FANUC 数控系统实验指导 …………………………………………… 213

Ⅲ.1　FANUC 数控系统 ………………………………………………………………… 213

Ⅲ.2　FANUC 数控车削编程 …………………………………………………………… 214

Ⅲ.3　FANUC 数控铣削编程 …………………………………………………………… 227

References ………………………………………………………………………………… 236

Chapter 1 Introduction of CNC

Objectives

- To understand the working principle of CNC machines.
- To understand the development of CNC systems.
- To understand the classifications of CNC machines.
- To understand the applications of CNC machines.

1.1 History of NC Development

Numerical Control Machine is called NC for short. It is an auto control technology which has been developed in modern times and a means by which the numerical information can fulfill the operation of the auto control machine. It minutes down in advance the machining procedure and the motion variable such as coordinate direction steering and speed of axes on the control medium in the form of numbers and it automatically controls the machine motion by the NC device at the same time. It also has some functions of finishing automatic tools conversion, automatic measuring, lubrication and automatic cooling etc.

1947 was the year in which Numerical Control was born. It began because of an urgent need. John C. Parsons of the Parson's Corporation, Michigan, a manufacturer of helicopter rotor blades could not make his templates fast enough, and then he invented a way of coupling computer equipment with a jig borer.

In 1949, US air force realized that parts for its planes and missiles were becoming more complex. Also the designs were constantly being improved; changes in drawings were frequently made. Thus in their search for methods of speeding up production, an air force study contract was given to the Parson's Corporation. The servomechanisms lab of MIT was the subcontractor.

Today the development of the NC machine completely depends on the NC system. The NC system has experienced two stages and six generations since American produced the first NC milling machine in 1952.

1. NC Stage (1952 – 1970)

The early computing speed was very low, which did not have too much effect on the scientific computing and the data handling. Man had to set up a machine specialized computer as a control system by using digital logic circuit, which was called Hard Wired NC, also NC for short. This stage experienced three generations.

The first generation of NC (1952 – 1959): Device was composed of electronic tube element.

The second generation of NC (1959 – 1965): Device was composed of transistor tube element.

The third generation of NC (1965 – 1970): Device was composed of small and medium scale integrated circuits were carried out.

2. CNC Stage (1970 –)

General-purpose, small-sized computers were mass-produced by 1970. Its computing speed was much higher than that in the 1950s and 1960s. These general-purpose, small-sized computers were much lower in cost and much higher in reliability than the specialized computers. Therefore, they were transferred as the kernel parts of the NC system. Since then they have come into computer numerical control (CNC) stage. With the development of computer technology, this stage also experienced three generations.

The fourth generation of NC (1970 – 1974): In this period, the small-sized, general-purpose computer control system of the large scale integrated circuit was greatly applied.

The fifth generation of NC (1974 – 1990): In this period the microprocessor was applied to the NC system.

The sixth generation of NC (1990 –): The personal computer (PC) performance has been developed greatly since the 1990s and it can meet the requirements of the kernel parts of the NC system. Since then the NC system has entered the PC-based era.

1.2 Concept of NC and CNC

1.2.1 NC Technology

1. Numerical Control

Numerical control (NC) is a form of programmable automation in which mechanical actions of a machine tool or other equipment are controlled by a program containing coded alphanumeric data. The alphanumerical data represent relative positions between a workhead and a work part as well as other instructions needed to operate the machine. The workhead is a cutting tool or other processing apparatus, and the workpiece is the object being processed. When the current job is completed, the program of instructions can be changed to process a new job. The capability to change the program makes NC suitable for low and medium productions. It is much easier to write new programs than to make major alterations of the processing equipment.

2. Basic Components of NC Machine Tools

The control system of a numerically controlled machine tool can handle many tasks commonly by the operator of a conventional machine. For this, the numerical control system must "know" when and in what sequence it should issue commands to change tools, at what speeds and feeds the machine tool should operate, and how to work a part to the required size. The system gains the ability to perform the control functions through the numerical input information—the control program, which is also called Part Program.

A typical NC machine tool has five fundamental units: the input media, the machine control unit, the servo-drive unit, the feedback transducer, and the mechanical machine tool unit (Figure 1.1).

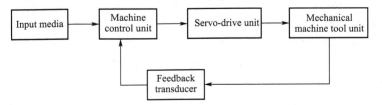

Figure 1.1 Basic Components of NC Machine Tool

The work process of NC is shown in Figure 1.2. The part programmer should study the part drawing and the process chart, and then prepare the control program on a standard form in the specified format. It contains all the necessary control information. A computer-assisted NC part programming for NC machining method is also available, in which the computer considerably facilitates the work of the programmer and generates a set of NC instructions. Next the part program is transferred to the control computer. The wide accepted method is that the worker types the part program into the computer from the keyboard of the computer numerical control front panel. The computer converts each command into a signal that the servo-drive unit needs. The servo-drive unit then drives the machine tool to manufacture the finished part.

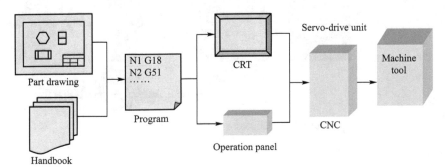

Figure 1.2 The Work Process of NC

1.2.2 CNC Technology

1. CNC system

In 1970s, computer numerically controlled (CNC) machine tools were developed with minicomputers being used as control units. With the advances in electronics and computer technology, current CNC systems employ several high-performance microprocessors and programmable logical controllers that work in a parallel and coordinated fashion.

Now, both NC and CNC mean Numerical Controller and there is no difference between them. Therefore, NC machine means a machine tool with a CNC system.

A CNC machine is an NC machine with the added feature of an onboard computer. The onboard computer is often referred to as the machine control unit (MCU). Control units for NC machines are usually hardwired, which means that all machine functions are controlled by the physical electronic elements that are built into the controller. The onboard computer, on the other hand, is "soft" wired, which means the machine functions are encoded into the computer at the time of manufacture, and they will not be erased when the CNC machine is turned off. Computer memory that holds such information is known as read-only memory (ROM).

The MCU usually has an alphanumeric keyboard for direct or manual data input (MDI) of part programs. Such programs are stored in random-access memory (RAM) portion of the computer. They can be played back, edited, and processed by the control. All programs residing in RAM, however, are lost when the CNC machine is turned off. These programs can be saved on auxiliary storage devices such as punched tape, magnetic tape, or magnetic disk. New MCU units have graphics screens that can display not only the CNC program but the cutter paths generated and any errors in the program.

2. Components of CNC System

CNC machine is composed of the following parts (Figure 1.3).

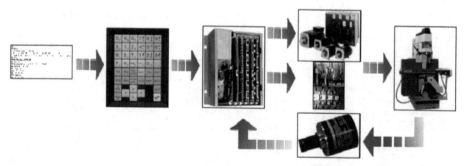

Figure 1.3 Structure of CNC Machine Tools

1) CNC device

The CNC device is the kernel of the CNC system. Its function is to handle the input part machining program or operation command. Then output control commands to the appropriate executive parts and finishes the work which the parts machining program and operation need. It mainly consists of computer system, position control panel, PLC interface panel, communication interface panel, extension function template and appropriate control software. The display unit serves as an interactive device between the machine and the operator. When the machine is running, the display unit displays the present status such as the position of the machine slide, the spindle revolutions per minute (RPM), the feedrate, the part program, etc.

In an advanced CNC machine, the display unit can show the graphics simulation of the tool path so that part program can be verified before the actually machining. Other important information about the CNC system can also be displayed for maintenance and installation work such as machine parameters, logic diagram of the programmer controller, error massages and diagnostic data.

2) Servo unit, drive device and measure device

Servo unit and drive device include spindle servo drive device, spindle motor, feed servo drive device and feed motor. Measure device means the position and speed measure device. It is a necessary device to finish the spindle control, closed-loop for the feed speed and for the feed position. Spindle servo can complete the cutting motion for the part machining and control the speed. The feed servo system can finish the shaping motion which the part machining need and control speed and position. The characteristic is to sensitively and accurately find the position of the CNC device and the speed command.

3) Control panel

Control panel, also called operation panel, is a tool used for mutual information between the operator and the CNC machine. The operator can operate, program and debug the CNC machine or set and alter the machine parameter. The operator can also understand and inquire the motion condition of the NC machine by using the control panel. It is input and output parts.

4) Control medium and program input and output equipment

The control medium is an agent to record the part machining program and it is also a medium to set up contraction between man and machine. Program input and output equipment are the devices by which the information exchange can be done between the CNC system and external equipment. It inputs the part machining program recorded on the control medium into the CNC system and stores or records the debugged part machining program on the appropriate medium with the output device. Today the control medium of the CNC machine and the program of input and output equipment are the disk and disk driver.

5) Machine body

The machine body of CNC system is an executive part to fulfill the machining parts. It is composed of the main motion parts, feed motion parts, bearing rack and special device, automatic platform change system, automatic tool changer (ATC) system and accessory device.

3. Input Method

The input media contains the program of instructions, Which include detailed step-by-step commands that direct the actions of the machine tool. The program of instructions is called a part program. The individual commands refer to positions of a cutting tool relative to the worktable on which the workpiece is fixed. Additional instructions are usually included, such as spindle speed, feed rate, cutting tool selection, and other functions. The program is coded on a suitable medium for submission to the machine control unit. For many years, the common medium was 1-inch wide punched tape, using a standard format that could be interpreted by the machine control unit. Today, punched tape has largely been replaced by newer storage technologies in modern machine shops. These technologies include magnetic tape, diskette, and electronic transfer of part programs from a computer.

1) Floppy disk drive

Floppy disk is a small magnetic storage device for CNC data input. It has been the most common storage media since the 1970s in terms of data transfer speed, reliability, storage size, data handling and the ability to read and write. Furthermore, the data within a floppy could be easily edited at any point as long as the operator has the proper program to read it. However, this method has proven to be quite problematic in the long run as floppies that have a tendency to degrade alarmingly fast and are sensitive to large magnetic fields and as well as the dust and scratches that usually existed on the shop floor.

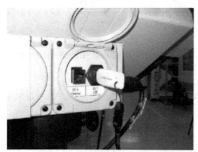

Figure 1.4 USB Flash Drive of CNC Machine

2) USB flash drive

A USB flash drive (Figure 1.4) is a removable and rewritable portable hard drive with compact size and bigger storage size than a floppy disk. Data stored inside the flash drive the controller. Once the downloaded section is executed, the section will be discarded to leave room for other sections. This method is commonly used for machine tools that do not have enough memory or storage buffer for large CNC part programs.

3) Serial communication

The data transfer between a computer and a CNC machine tool is often accomplished through a serial communication port (Figure 1.5). International standards for serial communications are established so that information can be exchanged in an orderly way. The most common interface between computers and CNC machine tools is referred to the EIA (Electronic Industries Association) standard RS-232. Most of the personal computers and CNC machine tools have built in RS-232 port and a standard RS-232 cable is used to connect a CNC machine to a computer, which enables the data transfer in reliable way. Part programs can be downloaded into the memory of a machine tool or uploaded to the computer for temporary storage by running a communication program on the computer and setting up the machine control to interact with the communication software.

Direct Numerical Control (DNC) is referred to a system connecting a set of numerically controlled machines to a common memory for part program or machine program storage with provision for on-demand distribution of data to the machines. The NC part programs are downloaded a block or a section at a time into the controller. Once the downloaded section is executed, the section will be discarded to leave room for other sec-

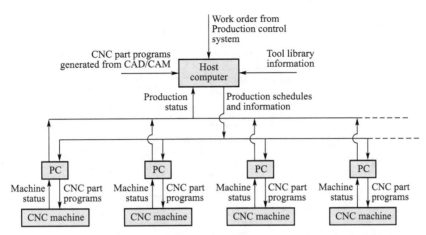

Figure 1.5 Serial Communication in a Distributed Numerical Control System

tions. This method is commonly used for machine tools that do not have enough memory or storage buffer for large NC part programs.

DNC is a hierarchical system for distributing data between a production management computer and NC systems (ISO 2806: 1994). The host computer is linked with a number of CNC machines or computers connecting to the CNC machines for downloading part programs. The communication program in the host computer can utilize two-way data transfer features for production data communication including: production schedule, parts produced and machine utilization etc.

4) Ethernet communication

Due to the advancement of the computer technology and the drastic reduction of the cost of the computer, it is becoming more practical and economical to transfer part programs between computers and CNC machines via an Ethernet communication cable. This medium provides a more efficient and reliable means in part programs transmission and storage. Most companies now have built a Local Area Network (LAN) as their infrastructure. More and more CNC machine tools provide an option of the Ethernet Card for direct communication within the LAN.

5) Conversational programming

Part program can be input to the controller via the keyboard. Built-in intelligent software inside the controller enables the operator to enter the required data step by step. This is a very efficient way for preparing program for relatively simple workpieces involving up to 2.5 axis machining.

4. CNC Manufacturing Process

The main stages involved in producing a component on a CNC system are shown in Figure 1.6.

(1) A part program is written through using G and M codes. It describes the sequence of operations that the machine must perform in order to manufacture the component.

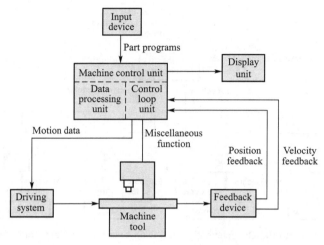

Figure 1.6　Working Principles of CNC Machines

This program can be produced off-line, i.e, away from the machine, either manually or with the aid of a CAD/CAM system.

(2) The part program is loaded into the machine computer, called the controller. At this stage, the program can still be edited or simulated using the machine controller keypad/input device.

(3) The machine controller processes the part program and sends signals to the machine components directing the machine through the required sequence of operations necessary to manufacture the component.

The application of CNC to a manual machine allows its operation to become fully automated. Combining the CNC with the use of a part program, the machine performs repeat tasks with high degrees of accuracy.

1.3　Classifications of CNC Machines

CNC machines are classified in different ways as follows:
(1) Types of CNC machines application.
(2) Types of CNC motion control system.
(3) Types of servo-drive system.

1.3.1　Types of CNC Machines Application

CNC machines are widely used in the metal cutting industry. They are best used to produce the following types of product:
(1) Parts with complicated contours.
(2) Parts requiring close tolerance and/or good repeatability.
(3) Parts requiring expensive jigs and fixtures if produced on conventional machines.

(4) Parts that may have several engineering changes, such as during the development stage of a prototype.

(5) In cases where human errors could be extremely costly.

(6) Parts that are needed in a hurry.

(7) Small batch lots or short production runs.

Some common types of CNC machines and instruments used in industry are as follows (Figure 1.7—Figure 1.11):

(1) Drilling machine.

(2) Lathe/Turning center.

(3) Milling/Machining center.

(4) Turret press and punching machine.

(5) Wire cut electro discharge machine (EDM).

(6) Grinding machine.

(7) Flame and Laser-cutting machines.

(8) Water jet cutting machine.

(9) Electro chemical machine.

(10) Coordinate measuring machine.

(11) Industrial robot.

Figure 1.7　Numerical Control Benders

Figure 1.8　Laser-beam Cutters

Figure 1.9　Electric Discharge Machines

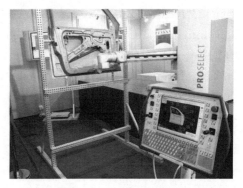

Figure 1.10　Laser Measuring Machines

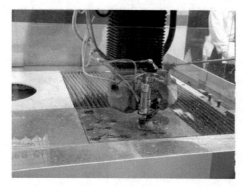

Figure 1.11 Super-high Pressure Water Jets

1.3.2 Types of CNC Motion Control System

Some CNC processes are performed at discrete locations on the workpiece (e.g. drilling, punching and spot welding). Others are carried out while the workhead is moving (e.g. turning, milling and continuous arc welding). If the tool is moving, it may be required to follow a straight-line path or a circular or other curvilinear path. These different types of movement are accomplished by the motion control system.

Motion control systems for CNC can be divided into three types: point-to-point control system, straight-cut control system and contouring control system.

1. Point-to-point Control System

Point-to-point control system, also called the Positing Control System, moves the worktable to a programmed location without regard for the path taken to get to the location. Once the move has been completed, some processing actions are accomplished by the workhead at the location, such as drilling or punching a hole. Thus, the program consists of a series of point locations at which operations are performed (Figure 1.12).

The machine control unit in a point-to-point control system contains registers that hold the individual axis motion commands. In some systems, the X-axis command is satisfied initially, followed by Y-axis and Z-axis commands. This operation may produce a zigzag path that will ultimately terminate at the proper point location.

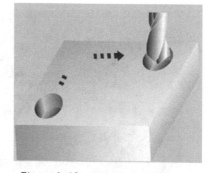

Figure 1.12 Point-to-point Control

2. Straight-cut Control System

Many CNC straight-cut systems contain a more complex MCU. In these servos, positioning commands are evaluated simultaneously so that vector motion in two axes is possible (Figure 1.13). However, the vector motion is limited to a one-to-one pulse output. Therefore, only 45° vectors maybe traced. Such systems are sometimes called the Straight-cut Control System.

3. Contouring Control System

The contouring facility enables a CNC machine to follow any path at any prescribed feedrate. The contouring control system, also called the Continuous Path Control System, manages the simultaneous motion of the cutting tool in two, three, four, or five axes (the fourth and fifth axes are angular orientations) by interpolating the proper path between prescribed points (Figure 1.14).

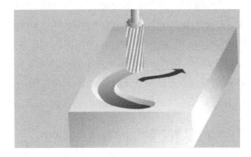

Figure 1.13 Straight-cut Control   Figure 1.14 Contouring Path Control

When continuous path control is utilized to move the tool parallel to only one of the major axes of the machine tool worktable, this is called Straight-cut CNC. When continuous path control is used for simultaneous control of two or more axes in machine operations, the term contouring is used. All CNC contouring systems have the ability to perform linear interpolation and circular interpolation.

4. Interpolation

One of the important aspects of contouring is interpolation. The paths that a contouring CNC system is required to generate often consist of circular arcs and other smooth nonlinear shapes. Some of these shapes can be defined mathematically by relatively simple geometric formulas, whereas others cannot be mathematically defined except by approximation. In any case, a fundamental problem in generating these shapes using CNC equipment is that they are continuous, whereas CNC is digital. To cut along a circular path, the circle must be divided into a serious of straight-line segments that approximate the circular path. The tool is commanded to machine each line segment in succession. So that the machined surface closely matches the desired shape. The maxim error between the nominal (desired) surface and the actual (machined) surface can be controlled by the lengths of the individual line segments.

If the programmer were required to specify the endpoints for each of the line segments, the programming task would be extremely arduous and fraught with errors. Also, the part program would be extremely long because of a large number of points. To ease the burden, interpolation routines have been developed that calculate the intermediate points to be followed by the cutter to generate a particular mathematically defined or approximated path. A number of interpolation methods are available to deal with the various problems encountered in generating a smooth continuous path. They include:

(1) Linear interpolation.
(2) Circular interpolation.
(3) Helical interpolation.
(4) Parabolic interpolation.

(5) Cubic interpolation.

Each of these procedures permits the programmer to generate machine instructions for linear or curvilinear paths using relatively few input parameters. The interpolation module in the MCU performs the calculation and directs the tool along the path. In CNC system, the interpolation is generally accomplished by software. Linear and circular interpolations are almost always included in modern CNC systems, whereas helical interpolation is a common option. Parabolic and cubic interpolations are less common. They are only needed by machine shops that must produce complex surface contours.

1.3.3 Types of CNC Servo-drive System

As the actual velocity and position detected from a sensor are fed back to a control circuit, the servo motor used in the CNC machine is continuously controlled to minimize the velocity error or the position error (Figure 1.15). The feedback control system consists of three independent control loops for each axis of the machine tool: the outermost control loop is a position-control loop, the middle loop is a velocity-control loop, and the innermost loop is a current-control loop. In general, the position-control loop is located in the CNC and the others are located in a servo driving device. However, there is no absolute standard about the location of control loops and the locations can be varied based on the intention of the designer.

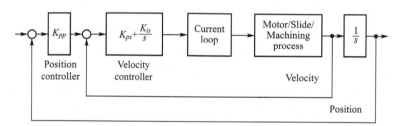

Figure 1.15 Three Kinds of Control Loop in CNC

In the spindle system of machine tools, feedback control of velocity is applied to maintain a regular rotation speed. The feedback signal is generally generated in two ways: a tacho-generator, which generates an induction voltage (analog signal) as a feedback signal, and an optical encoder, which generates pulses (digital signals). In recent times, it is typical that feedback control is performed based on an optical encoder signal instead of a tachometer signal.

1. Open-loop Servo-drive System

Open-loop system has no access to the real time data about the performance of the system and therefore no immediate corrective action can be taken in case of system disturbance. This system is normally applied only to the case where the output is almost constant and predictable. Therefore, an open-loop system is unlikely to be used to control machine tools since the cutting force and loading of a machine tool is never a constant. The only exception is the wire cut machine for which some machine tool builders still prefer to use an open-loop system because there is virtually no cutting force in wire cut machining.

2. Semi-closed-loop Servo-drive System

A semi-closed-loop control CNC system uses feedback measurements to ensure that the worktable is moved to the desired position. It is characterized as a system that the indirect feedback monitors the output of servomotor. Although this method is popular with CNC systems, it is not as accurate as direct feedback. The half-closed-loop system compares the command position signal with the drive signal of the servomotor.

In operation, the half-closed-loop system is directed to move the worktable to a specified location as defined by a coordinate value in a Cartesian system. Most positioning systems have at least two axes with a control system for each axis, but our diagram only illustrates one of these axes. A servomotor connected to a leadscrew is a common actuator for each axis. A signal indicating the coordinate value is sent from the MCU to the motor that drive the leadscrew, whose rotation is converted into linear motion of positioning table. As the table moves closer to the desired coordinate value, the difference between the actual position and the input value is reduced. The actual position is measured by a feedback sensor, which is attached to servomotor axis or lead screw. This system is unable to sense backlash or lead screw windup due to varying loads, but it is convenient to adjust and has a good stability.

The semi-closed-loop is the most popular control mechanism. In this type, a position detector is attached to the shaft of a servo motor and detects the rotation angle. The position accuracy of the axis has a great influence on the accuracy of the ball screw. For this reason, ball screws with high accuracy were developed and are widely used. Due to the precision ball screw, the problem with accuracy has been overcome practically.

If necessary, pitch-error compensation and backlash compensation can be used in CNC to increase the positional accuracy. The pitch-error compensation method is that, at the specific pitch, the instructions to the servo driver system are modified in order to remove the accumulation of positional error. The backlash compensation method is that, whenever the moving direction is changed, additional pulses corresponding to the amount of backlash are sent to the servo driver system. Recently, the usage of the Hi-Lead-type ball screw with large pitch for high-speed machining has increased.

3. Closed-loop Servo-drive System (Figure 1.16)

In a closed-loop system, feed back devices closely monitor the output and any disturbance will be corrected in the first instance. Therefore, high system accuracy is achievable. This system is more powerful than the open-loop system and can be applied to the case where the output is subjected to frequent change. Nowadays, almost all CNC machines use this control system.

A half closed-loop or closed-loop system uses conventional variable-speed AC or DC motors, called servos, to drive the axes.

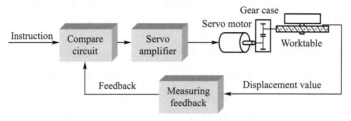

Figure 1.16 Closed-loop Servo-drive System

1.4 CNC Application

1.4.1 Advantages and Disadvantages

1. NC

The advantages generally attributed to NC, with emphasis on machine tool applications, are the following:

(1) Greater accuracy and repeatability. Compared with manual production methods, NC reduces or eliminates variations that are due to operator skill differences, fatigue, and other factors attributed to inherent human variability. Parts are made closer to nominal dimensions, and there is less dimensional variation among parts in the batch.

(2) More complex part geometries are possible. NC technology has extended the range of possible part geometries beyond what is practical with manual machining methods. This is an advantage in product design in several ways.

More functional features can be designed into a single part to reduce the total number of parts in the product and the associated cost of assembly. Mathematically defined surfaces can be fabricated with high precision. The space is expanded within which the designer's imagination can wander to create new part and product geometries.

(3) Nonproductive time is reduced. NC cannot optimize the metal cutting process itself, but it does increase the proportion of time. The machine is cutting metal, reducing the workpiece handling time, and carrying out automatic tool changes on some NC machines. This advantage leads to labor cost savings and lowers elapsed times to produce parts.

(4) Lower scrap rates. Because greater accuracy and repeatability are achieved, and human errors are reduced during production, more parts are produced within tolerance. As a consequence, a lower scrap allowance can be planned into the production schedule, so fewer darts are made in each batch with the result that production time is saved.

(5) Inspection requirements are reduced. Less inspection are needed when NC is used because parts produced from the same NC part programs are virtually identical. Once the program has been verified, there is no need for the high level of sampling inspection that is required when parts are produced by conventional manual methods. Except for tool wear

and equipment malfunctions, NC produces exact replicates of the part in each cycle.

(6) Engineering changes can be accommodated more gracefully. Instead of making alterations in a complex fixture so that the part can be machined to the engineering change, revisions are made in the NC part program to accommodate the change.

(7) Simpler fixtures are needed. NC requires simpler fixtures because accurate positioning of the tool is accomplished by the NC machine tool. Tool positioning does not have to be designed into the jig.

(8) Shorter manufacturing lead times. Jobs can be set up more quickly and fewer setups are required per part when NC is used. This results in shorter elapsed time between order release and completion.

(9) Reduced parts inventory. Because fewer setups are required and jig changeovers are easier and faster, NC permits production of parts in smaller lot sizes. The economic lot size is lower in NC than in conventional batch production. Average parts inventory is therefore reduced.

(10) Less floor space required. This result from the fact that fewer NC machines are required to perform the same amount of work compared with the number of conventional machine tools needed. Reduced parts inventory also contributes to lower floor space requirements.

(11) Operator skill-level requirements are reduced. The skill requirements for operating an NC machine are generally less than those required to operate a conventional machine tool. Tending an NC machine usually consists only of loading and unloading parts and periodically changing tools. The machining cycle is carried out under program control. Tools changing for some NC machine tools can even be carried out by program control. Performing a comparable machining cycle in a conventional machine requires much more participation by the operator, and a higher level of training and skill are needed.

Contrarily, the disadvantages of NC include the following:

(1) Higher investment cost. An NC machine tool has a higher first cost than a comparable conventional machine tool for the following reasons:

① NC machines include CNC controls and electronics hardware.

② Software development costs of the CNC controls and manufacturer must be included in the cost of the machine.

③ More reliable mechanical components are generally used in NC machine.

④ NC machine tools often possess additional features not included on conventional machines, such as automatic tool changers and part changers.

(2) Higher maintenance effort. In general, NC equipment requires a higher level of maintenance than conventional equipment, which means higher maintenance and repair costs. This is due largely to the computer and other electronics that are included in modern NC system. The maintenance staff must include the persons who are trained in maintaining and repairing this type of equipment.

(3) Part programming. NC equipment must be programmed. To be fair, it should be mentioned that process planning must be accomplished for any part, whether or not it is

produced on NC equipment. However, NC part programming is a special preparation step in batch production that is absent in conventional machine shop operations.

(4) Higher utilization of NC equipment. To maximize the economic benefits, an NC machine tool usually must be operated multiple shifts. This might mean adding one or two extra shifts to the plant's normal operations, with the requirement for supervision and other staff support.

2. CNC

CNC opens up new possibilities and advantages not offered by older NC machines.

(1) Reduction in the hardware necessary to add a machine function. New functions can be programmed into the MCU as software.

(2) The CNC program can be written, stored, and executed directly at the CNC machine.

(3) Any portion of an entered CNC program can be played back and edited at will. Tool motions can be electronically displayed upon playback.

(4) Many different CNC programs can be stored in the MCU.

(5) Several CNC machines can be linked together to a main computer. Programs written via the main computer can be downloaded to any CNC machine in the network. This is known as DNC.

(6) Several DNC systems can also be networked to form a large distributive numerical control system.

(7) The CNC program can be input from flash or floppy disks or downloaded from local area networks.

CNC machines can dramatically boost productivity. The CNC manager, however, can only ensure such gains by first addressing several critical issues as the following:

(1) Sufficient capital must be allocated for purchasing quality CNC equipment.

(2) CNC equipment must be maintained on a regular basis by obtaining a full-service contract or by hiring an in-house technician.

(3) Personnel must be thoroughly trained in the operation of CNC machines. In particular, many jobs require setups for machining parts to comply with tolerances of form and function.

(4) Careful production planning must be studied because the hourly cost of operation of a CNC machine is usually higher than that for conventional machines.

1.4.2 Financial Rewards of CNC Investment

Investors are encouraged to look to the CNC machine tool as a production solution with the following benefits:

(1) Savings in direct labor. One CNC machine's output is commonly equivalent to that of several conventional machines.

(2) Savings in operator training expenses.

(3) Savings in shop supervisory costs.

(4) Savings due to tighter, more predictable production scheduling.

(5) Savings in real estate, since fewer CNC machines are needed.

(6) Savings in power consumption, since CNC machines produce parts with a minimum of motor idle time.

(7) Savings from improved cost estimation and pricing.

(8) Savings due to the elimination of construction of precision jigs, reduced need for special fixtures, and reduced maintenance and storage costs of these items.

(9) Savings in tool engineering/design and documentation. The CNC machining capability eliminates the need for special form tools, special boring bars, special thread cutters, etc.

(10) Reduced inspection time due to the CNC machine's ability to produce parts with superior accuracy and repeatability. In many cases, only spot-checking of critical areas is necessary without loss of machine time.

Payback period is used to estimate investment efficiency. The payback period calculation estimates the number of years required to recover the net cost of the CNC machine tool.

$$\text{Payback Period} = \frac{\text{Net Cost of CNC} - \text{Net Cost of CNC} \times \text{Tax Credit}}{\text{Savings} - \text{Savings} \times \text{Tax Rate} + \text{Yearly Depreciation of CNC} \times \text{Tax Rate}} \quad (1.1)$$

Return on investment (ROI) is used to estimate investment efficiency. The ROI calculation predicts the percent of the net cost about CNC which will be recovered each year. The ROI calculation accounts for the useful life of the CNC machine tool.

$$\text{ROI} = \frac{\text{Average Yearly Savings} - \text{Net Cost of CNC/Year of Life}}{\text{Net Cost of CNC}} \quad (1.2)$$

Example 1:

Given the investment example (Table 1.1) for implementing a new CNC machine tool, determine the payback period and the annual return on investment. The CNC is conservatively estimated to have a useful life of 12 years.

Table 1.1 Financial Rewards of CNC Investment

Initial investment ($)	130,250
One-time savings in tooling ($)	35,000
Net cost of CNC ($)	95,250
Average yearly savings ($)	63,100
Tax credit (10%)	0.1
Tax rate (46%)	0.46
Yearly depreciation of CNC ($)	10,900

$$\text{Payback Period} = \frac{95,250 - 95,250 \times 0.1}{63,100 - 63,100 \times 0.46 + 10,900 \times 0.46} \quad (1.3)$$

$$\text{Payback Period} = 2.19 \text{ years}$$

This calculation estimates that the net cost of the CNC will be recovered in 2.19 years.

$$\text{ROI} = \frac{63,100 - 95,250/12}{95,250} \qquad (1.4)$$

$$\text{ROI} = 0.57$$

This calculation estimates that the investor can expect 57% of the net cost of the CNC (or $0.57 \times \$95,250 = \$54,293$) to be recovered each year if the CNC machine's useful life is 12 years.

1.4.3 Reliability of CNC Machines

Reliability is defined as the probability that a device will perform its required function under stated conditions for a specific period of time. The life of a population of units can be divided into three distinct periods. Figure 1.17 shows the reliability "bathtub curve" which models the instantaneous failure rates vs. time. This first period is also called infant mortality period. The next period is called the useful life (normal Life). The third period begins at the point where the slope begins to increase and extends to the end of the graph. This is what happens when units become old and begin to fail at an increasing rate.

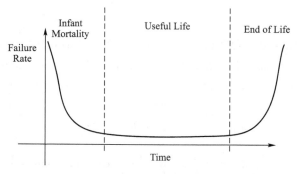

Figure1.17 The Bathtub Curve

Mean Time between Failure (MTBF) is a reliability term used to provide the amount of failures per million hours for a product. This is the most common inquiry about a product's life span, and is important in the decision-making process of the end user. MTBF is the predicted elapsed time between inherent failures of a system during operation. MTBF can be calculated as the arithmetic mean (average) time between failures of a system.

$$\text{MTBF} = T/R \qquad (1.5)$$

T——Total time.

R——Number of failures.

Mean Time to Repair (MTTR) is the time needed to repair a failed hardware module. In an operational system, repair generally means replacing a failed hardware part. Thus, hardware MTTR could be viewed as mean time to replace a failed hardware module. Taking too long to repair a product drives up the cost of the installation in the long run. To avoid MTTR, many companies purchase spare products so that a replacement can be installed quickly.

MTTR is a basic measure of the maintainability of repairable items. It represents the average time required to repair a failed component or device. Expressed mathematically, it is the total corrective maintenance time divided by the total number of corrective maintenance actions during a given period of time.

$$MTTR = T/N \tag{1.6}$$

T——Total corrective maintenance time.

N——Number of units under test.

Exercises

1. Narrate the concept of NC.
2. Narrate the concept of CNC.
3. What are the components of CNC machine tool? What is the function of each component?
4. What is point-to-point control? What is contouring control?
5. What is open-loop control, half-closed-loop control and closed-loop control?
6. What are the advantages and disadvantages of CNC?
7. Narrate the financial rewards of CNC investment using return on investment.

Chapter 2 CNC Part Programming

Objectives

- To understand the structure of a CNC part programming.
- To master G-codes and other functions of CNC part programming.
- To understand the characteristics of new CAM technology.
- To understand the flow of CAM system.

2.1 Introduction

Part program is a list of coded instructions which describes how the designed component, or part, will be manufactured. These coded instructions are called data, a series of letters and numbers. The part program includes all the geometrical and technological data to perform the required machine functions and movements to manufacture the part.

The part program can be further broken down into separating lines of data, and each line describes a particular set of machining operations. These lines, which run in sequence, are called blocks. A block of data contains words which are sometimes called codes. Each word refers to a specific cutting/movement command or machine function.

The programming language recognized by the CNC, the machine controller, is an ISO code, which includes the G and M code groups. Each program word is composed from a letter, called the address, along with a number.

The part program can contain a number of separate programs, which together describe all the operations required to manufacture the part. The main program is the controlling program, i.e, the program first read, or accessed, when the entire part program sequence is run. This controlling program can call a number of smaller programs into operation.

These smaller programs, called sub programs, are generally used to perform repeating tasks, before returning control back to the main program. Normally, the controller operates according to one program. In this case, the main program is also the part program. Main programs are written by using ISO address codes listed below.

The program of instructions is the detailed step-by-step commands that direct the actions of the processing equipment. In machine tool applications, the program of instructions is called a part program, and the person who prepares the program is called a part programmer. CNC is a form of automatically operating a machine tool based on coded alphanumeric data. A complete set of coded instructions for executing an operation is called a program. The program is translated into corresponding electrical signals for input to servo motors that run the machine.

We call the whole process, from part graphics to finishing control medium, the programming of computer numerical control manufacturing, or the CNC programming. When using numerical control machine tool to manufacture parts, the programming is very important. The program is not only correct and fast but also effective and economical.

2.1.1 The Contents and Steps of CNC Programming

Before NC programming, the programmer should understand the numerical control machine specifications, characteristics, the functions and programming instruction format of the CNC system, etc. When programming, he should analyze the part's technical requirements, geometrical shape dimensions and technological requirements. Then he can determine the manufacturing method and calculate numerical value, get cutter position. According to part dimension, cutter position value, cutting parameters (spindle speed, feedrate, cutting depth) and auxiliary functions (ATC, CW, CCW, coolant on and off), the programmer can program. The program can be inputted into CNC system and the CNC system controls CNC machine tools to manufacture automatically.

Typically, CNC programmer follows a certain processor workflow that can be summarized into a several critical points or steps: analyzing part graphics, determining the manufacturing technological process, calculating numerical value, programming, verifying the program and inputting the program into CNC system.

1) Analyzing part graphics and determining the manufacturing technological process

This step includes analyzing the part graphics, understanding the machining contents and requirements, determining technological processes, machining plans, machining sequence, machining routes, fixing methods, cutting parameters and selecting suitable cutting tools, etc. Besides, the numerical control machine codes should be understood clearly and the numerical control machine functions should be exploited fully.

2) Correctly selecting program origin and coordinate system

In numerical control machine tools manufacturing, correctly selecting program origin and coordinate system is very important. On CNC programming, the program coordinate system is the standard coordinate system ascertained on the workpiece.

3) Calculating numerical values

After finishing the technological process, the next step is to get the tool path according to the part geometric dimension and the method of cutter radius compensation. Thus, we will obtain the cutter position.

4) Writing part programs

After determining machining route, technological process and the coordinate value of tool path, step-by-step, the programmer can write the program in accordance with the specified function codes and program formats of CNC system.

5) Verifying part programs

Before the program is used in real production, we must check the program. We detect tool path errors that could ruin the part, damage the fixtures, break the cutting tool or crash the machine, etc. In some cases, we test the program through manufacturing a part on a machine. On the basis of detecting result, the program is needed to be modified and to be adjusted until the program satisfies the machining requirements completely.

6) Completing documentation and sending program to machine shop

It is an unpleasant reality that many CNC programs, regardless of how they were developed, lack any background information that can help the machine operator. At best, the part program may include some basic data regarding the setup and even some special instructions. The operator needs to know what fixture has been used, how the part is oriented, what tools have been selected, and where the part is located.

The program is coded on a suitable medium for submission to the machine control unit. For many years, the common medium was 1-inch wide punched tape using a standard format that could be interpreted by the machine control unit. Today, punched tape has largely been replaced by newer storage technologies in modern shops. These technologies include magnetic tapes, diskettes, USB flash and electronic transfer of part programs from a computer.

The steps above-mentioned are programmed manually. This programming method is called manual part programming. We also know that a programmer not only must have the knowledge of the structure of machine tools, the functions and standards of CNC system, but also must have the knowledge of technological process, such as fixtures, cutting tools and cutting parameters, and so on.

Keep in mind that this is not always a step-by-step method as it may appear to be. Often, a decision made in one step influences a decision made in another step, which often leads to revisiting earlier stages of the process and making necessary changes.

Once the drawing and program reach the machine shop, it is up to the CNC operator to continue with actual production. Production cannot start right away, and certain workflow is followed at the machine as well:

(1) Evaluate the part program.

(2) Check supplied material.

(3) Prepare required tools.

(4) Set up and register tools.

(5) Set up part in a fixture.

(6) Load program.

(7) Set various offsets.

(8) Run first part.

(9) Optimize program if necessary.

(10) Run production.

(11) Inspect frequently.

2.1.2 The Methods of CNC Programming

Part programming can be accomplished by using a variety of procedures ranging from manual method to highly automated method.

1. Manual Part Programming

The programming is called manual part programming which whole part programming is completed manually (including calculating numerical value on a computer).

In many mechanical manufacturing trades, there are a large number of uncomplicated parts that are constituted only by the simple geometric elements of straight lines and circles. The numerical values of the parts are calculated simply. The blocks of a program aren't numerous. And checking the program is easy. These part programs can be completed manually. So manual programming is still a very common programming method at home and abroad.

But manual programming has difficulty in programming complicated parts that have non-circular curves and surfaces. So we must use the automatic part programming to program.

The procedure for manual programming can be divided into four steps:

(1) Analysis of workshop drawings.

(2) Definition of work plans.

(3) Choice of clamping devices and necessary tools (set-up sheet).

(4) Generating the CNC program (program sheet).

Various documents must be analyzed and plans for production execution must be created (Figure 2.1).

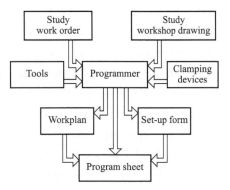

Figure 2.1 Procedure for Manual Programming

2. Automatic Programming

Automatic programming is also called computer-aided programming. Most or all of the programming is completed by a computer, such as calculating numerical values, writing programs, fabricating the control medium, etc. Automatic programming lightens programmer intensity of labor, shortens the programming time and improves the programming quality. At the same time, it solves the complicated part programming which is impossible to program by manual programming. The more complicated shape and technological process the parts have, the clearer the superiority of automatic programming is.

The automatic programming method can be classified into the language-type programming method (e.g. APT: Automatically Programmed Tools) and the conversational programming method. In the language-type programming method, the machining sequence, part shape, and tools are defined in a language that can be understood by humans. The human-understandable language is then converted into a series of CNC-understandable instructions. In the conversational programming method, the programmer inputs the data for the part shape interactively using a GUI (Graphical User Interface), selects machining sequences, and inputs the technology data for the machining operation. Finally, the CNC system generates the part program based on the programmer's input. Typically conversational programming can be carried out by an external CAM system and a symbolic conversational system that is located either inside the CNC system or in an external computer.

2.2　The Basis of CNC Part Programming

2.2.1　CNC Coordinate Systems

At any time, the location of CNC machine tools is controlled by a system of XYZ coordinates called Cartesian Coordinates. This system is composed of three directional lines, called axes, mutually intersecting at an angle of 90°. The point of intersection is known as the origin.

Almost everything that can be produced on a conventional machine tool can be produced on a computer numerical control machine tool. The machine tool movements used in producing a product are of two basic types: point-to-point (straight-line movements) and continuous path (contouring movements).

The Cartesian, or rectangular coordinate system was devised by the French mathematician and philosopher René Descartes. With this system, any specific point can be described in mathematical terms from one point along three perpendicular axes. This concept fits machine tools perfectly. Their construction is generally based on three axes of motion (X, Y, Z) plus an axis of rotation (Figure 2.2). On a plain vertical milling machine, the X axis is the horizontal movement (right or left) of the table, the Y axis is the table cross movement (toward or away from the column), and the Z axis is the vertical movement of the knee or the spindle. CNC systems heavily rely on the use of rectangular coordinates because the programmer can locate every point on a job precisely.

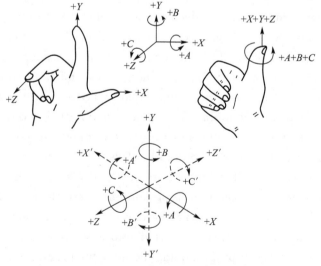

Figure 2.2　Coordinate System

Axis of motion is shown in Figure 2.3.

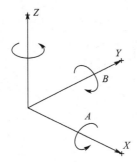

Figure 2.3 Axis of Motion

In generally, all motions have 6 degrees of freedom. In other words, motion can be resolved into 6 axes, namely, 3 linear axes (X, Y and Z axis) and 3 rotational axes (A, B, and C axis).

Controllable feed and rotation axis workpiece machining on CNC machine tools requires controllable and adjustable in feed axes which are run by the servo motors independently of each other. The hand wheels of typical conventional machine tools are consequently redundant in a modern machine tool.

CNC lathes (Figure 2.4) have at least 2 controllable or adjustable feed axes marked as X and Z. In CNC milling (Figure 2.5) the main function of the workpiece clamping devices is the correct positioning of the workpieces. The workpiece clamping should allow a workpiece change which is as quick, easy to approach, correctly and exactly positioned, reproducible as possible. For simple machining controllable, hydraulic chuck jaws are sufficient. For milling, on all sides, the complete machining should be possible with as few reclamping as possible. Machine/workpiece coordinate system are shown in Figure 2.6 and Figure 2.7.

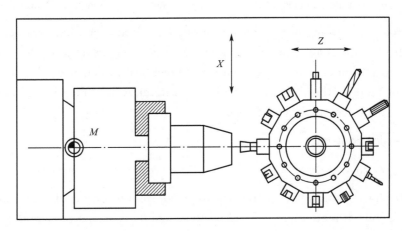

Figure 2.4 Controllable CNC Axes on an Automatic Lathe

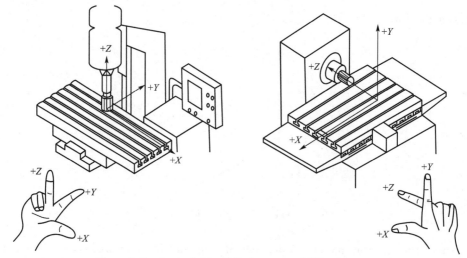

Figure 2.5　Controllable CNC Axes on a Milling Machine

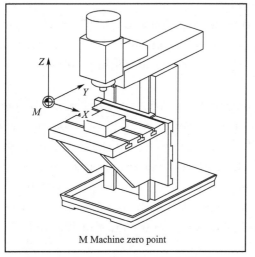

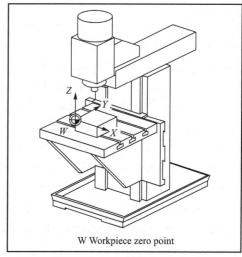

Figure 2.6　Machine Coordinate System　　　Figure 2.7　Workpiece Coordinate System

For complicated milling parts, with integrated automatic rotation, are being manufactured or built out of available modular systems to allow, as far as possible, complete machining without reclamping. Workpiece pallets, which are loaded with the next workpiece by the operator outside the work room and then automatically taken into the right machining position, are increasingly being used.

2.2.2　Dimension Systems

1. Absolute System

In an absolute system, all references are made to the origin of the coordinate system. All commands of motion are defined by the absolute coordinate referred to the origin.

2. Incremental System

This type of control is always used as a reference to the preceding point in a sequence of points. The disadvantage of this system is that if an error occurs, it will be accumulated.

2.2.3 Types of Zero Point and Reference Point

⊕	M	machine zero point
⊕	W	work part zero point
⊕	R	reference point
⊕	E	tool reference point
⊕	B	tool setup point
⊕	A	tool shank point
⊕	N	tool change point

1. Machine Zero Point M

Each numerically controlled machine tool works with a machine coordinate system. The machine zero point is the origin of the machine-referenced coordinate system (Figure 2.8). It is specified by the machine manufacturer and its position cannot be changed. In general, the machine

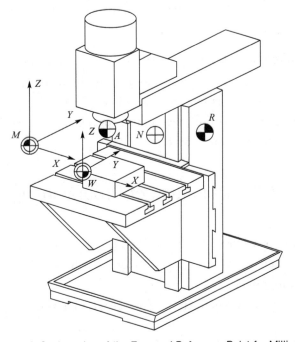

Figure 2.8 Location of the Zero and Reference Point for Milling

zero point M is located in the center of the work spindle nose for CNC lathes and above the left corner edge of the workpiece carrier for CNC vertical milling machines.

2. Workpiece Zero Point W (Program Zero Point)

The programming zero point is related to the origin position on the graphics. It is a logical reference point from which to work. So the programmer must decide where the origin of the coordinate axis system should be located. This decision is usually based on programming convenience and easiness. For example, the origin might be located at one of the corners of the part. If the workpiece is symmetrical, the zero point might be most conveniently defined at the center of symmetry. Wherever the location is, this zero is communicated to the machine tool operator. At the beginning of the job, the operator must move the cutting tool under manual control to some target point on the worktable, where the tool can be easily and accurately positioned. The part programmer has previously referenced the target point to the origin of the coordinate axis system. The operator inputs the coordinate value of zero point to MCU.

3. Reference Point R

A machine tool with an incremental travel path measuring system needs a calibration point which also serves for controlling the tool and workpiece movements. This calibration point is called the reference point R. Its location is set exactly by a limit switch on each travel axis. The coordinates of the reference point, with reference to the machine zero point, always have the same value. This value has a set adjustment in the CNC control. After switching the machine on the reference point, the machine tool has to be approached from all axes to calibrate the incremental travel path measuring system.

4. Tool Point

If you measure tools outside of the machine, then the reference points on the tool, tool holder (setup) and tool shank are of importance, because the control must reference the geometry information (such as length and radius) of a tool to a certain point in order to apply the coordinate values from the machining program to the workpiece precisely. Tool change should be in a safe location (point).

2.3 Definition of Programming

CNC programming is where all the machining data are compiled and where the data are translated into a language which can be understood by the control system of the machine tool. The machining data is as follows:

(1) Machining sequence classification of process, tool start up point, cutting depth, tool path etc.

(2) Cutting conditions spindle speed, feed rate, coolant, etc.

(3) Selection of cutting tools.

A CNC program consists of blocks, words and addresses (Figure 2.9 and Table 2.1).

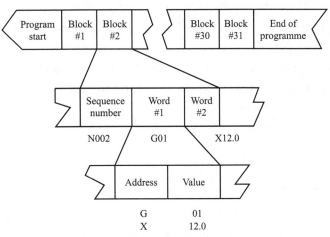

Figure 2.9 Structure of CNC Part Program

Table 2.1 Address

Address	Code/Syntax	Function
% MM	ISO	Subprogram
% PM	ISO	Program start main program
% TM	ISO	Tool compensation table
N	9001—9999999	Part program and subprogram number
N	1—8999	Block number
X	+/−9999,99	Distance/move in mm
Y	+/−9999,99	Distance/move in mm
Z	+/−9999,99	Distance/move in mm
B	+/−9999,99	Distance/move in degrees
R	+/−9999,99	Circle radii in mm
I	+/−9999,99	Circle center in X
J	+/−9999,99	Circle center in Y
K	+/−9999,99	Circle center in Z
P	0—99	Point definition
F	1—5000	Feed mm/rev or mm/min
S	20—99999	Spindle RPM
T	0—99	Tool offset number
E	0—99	Parameter in sub programs

(1) Block——A command given to the control unit is called block.

(2) Word——A block is composed of one or more words. A word is composed of an

identification letter and a series of numerals, e. g. , the command for a feed rate of 200mm/min is F200.

(3) Address——The identification letter at the beginning of each word is called address. Each block (program line) contains addresses which appear in this order.

N, G, X, Y, Z, F, S, T, M;

This order should be maintained throughout every block in the program, although individual blocks may not necessarily contain all these addresses.

An example of a program is as follows:

N20 G01 X20.5 F200 S1000 M03;
N21 G02 X30.0 Y40.0 I20.5 J32.0;

1. Sequence Number (N Address)

A sequence number is used to identify the block. It is always placed at the beginning of the block and can be regarded as the name of the block. The sequence numbers should not be consecutive. The execution sequence of the program is according to the actual sequence of the block, not the sequence of the number. In fact, some CNC systems do not require sequence numbers.

2. Preparatory Function (G Address)

A preparatory function determines how the tool is to move to the programmed target. The most common G addresses are listed in Table 2.2.

Table 2.2 G-codes (Lathe)

Standard G-code	Function
♯G00	Positioning (Rapid feed)
G01	Straight interpolation
G02	Circular interpolation (CW)
G03	Circular interpolation (CCW)
G04	Dwell
G20	Data input (inch)
♯G21	Data input (mm)
♯G22	Stored distance limit is effective (Spindle interference check ON) Stored
G23	distance limit is ineffective (Spindle interference check OFF)
G27	Machine reference return check
G28	Automatic reference return
G29	Return from reference
G30	2nd reference return
♯G32	Thread process
G40	Cancel of compensation
G41	Cutter compensation, Left
G42	Cutter compensation, Right

(续)

Standard G-code	Function
G50	Creation of virtual coordinate/Setting the rotating time of principal spindle
G70	Compound repeat cycle (Finishing cycle)
G71	Compound repeat cycle (Stock removal in turning)
G72	Compound repeat cycle (Stock removal in facing)
G73	Compound repeat cycle (Pattern repeating cycle)
G74	Compound repeat cycle (Peck drilling in Z direction)
G75	Compound repeat cycle (Grooving in X direction)
G76	Compound repeat cycle (Thread process cycle)
G90	Fixed cycle (Process cycle in turning)
G92	Fixed cycle (Thread process cycle)
G94	Fixed cycle (Facing process cycle)
G96	Control the circumference speed uniformly/Constant surface speed control (mm/min)
♯G97	Cancel the uniform control of circumference speed/Constant surface speed control cancel (rpm)
G98	Designate the feedrate per minute (mm/min)
♯G99	Designate the feedrate per the rotation of principal spindle (mm/rev.)
G90	Absolute programming
G91	Incremental programming

Note: 1. ♯ mark instruction is the modal indication of initial condition which is immediately available when power is supplied.
2. In general, the standard G-code is used in lathe, and it is possible to select the special G-code according to setting of parameters.

3. Parameter for Circular Interpolation (I/J/K Address)

These parameters specify the distance measured from the start point of the arc to the center. Numerals following I, J and K are the X, Y and Z components of the distance respectively.

4. Spindle Function (S Address)

The spindle speed is commanded under an S address and is always in revolution per minute. It can be calculated by the following formula:

$$\text{Spindle Speed} = \frac{\text{Surface Cutting Speed(m/min)} \times 1000}{\pi \times \text{Cutter Diameter(mm)}} \quad (2.1)$$

5. Feed Function (F Address)

The feed is programmed under an F address except for rapid traverse. The unit may be in mm per minute (in the case of milling machine) or in mm per revolution (in the case of turning machine). The unit of the feedrate has to be defined at the beginning of the program. The feed rate can be calculated by the following formula:

$$\text{Feed Rate} = \text{Chip Load} \times \text{No. of Tooth} \times \text{Spindle Speed} \quad (2.2)$$

Chip load——Feed per tooth.

6. Miscellaneous Function (M Address)

The miscellaneous function is programmed to control the machine operation rather than for coordinate movement. M codes that are especially useful in many programming applications are shown in Table 2.3.

Table 2.3 M Functions

M Function	Meaning
M00	Program stop
M02	End of program
M03	Spindle start (forward CW)
M04	Spindle start (reverse CCW)
M05	Spindle stop
M06	Manual/Automatic tool change with automatic travel to a fixed machine-specific change position
M07	Internal cooling lubricant supply ON
M08	External cooling lubricant ON
M09	Cooling lubricant supply OFF
M10	Chuck-clamping
M11	Chuck-unclamping
M12	Tailstock spindle out
M13	Tailstock spindle in
M19	Spindle stop with defined final position
M30	Program end with reset of CNC control system to ready condition
M66	Manual tool change in the position last moved to
M67	Tool data activation without tool change
M98	Transfer to subprogram
M99	End of subprogram

2.4 Part Programming

There are many codes included in a program. G codes perform preparatory functions and M codes perform auxiliary functions. They are the base of CNC programs. ISO (International Organization for Standardization) has worked out the standards of G codes and M codes. Because new CNC systems and machines have been emerging, a lot of functions in many systems surpass ISO standards. Their codes are abundant and their formats are flexible, which aren't restrained by ISO standards. In addition, even with the same function, the code and format are different among systems made by different companies. And codes and formats are also different between new and old systems made in the same company. But the preparatory functional codes and auxiliary functional codes in most of CNC systems are following ISO standards or similar standards.

Usually, several tools are used for machining one workpiece. The tools have different tool lengths (Figure 2.10). It is very troublesome to change the program in accordance with the tools.

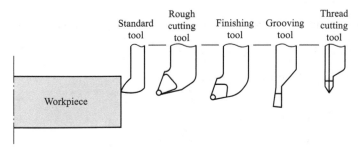

Figure 2.10 Tool Offset

Therefore, the length of each tool used should be measured in advance. By setting the difference between the length of the standard tool and the length of each tool in the CNC, machining can be performed without altering the program even when the tool is changed. This function is called Tool Length Compensation.

2.4.1 Turning Programming of MAHO GR350C

The front panel of MAHO GR350C is shown in Figure 2.11.

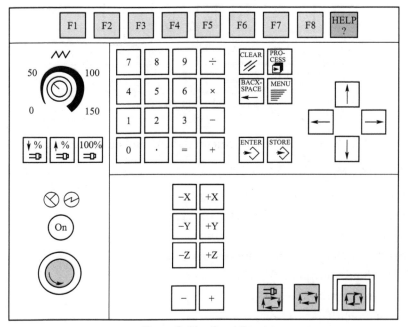

Figure 2.11 Front Panel

One part program (MAHO GR350C):

```
%PM1234567              (%PM+N   N-Program number)
N1234567                (Body)
N1 G18;
N2 G52;
...
N2000 M30;              (End)
```

1. G10 Axial Volume and Surface-orientated Pre-finishing Machining (Figure 2.12)

G10 X__(U__) Z__(W__) I__ K__ C__ (F__)

X——Start S: X (Absolute dimensioning).

U——Start S: X (Incremental dimensioning).

Z——Start S: Z (Absolute dimensioning).

W——Start S: Z (Incremental dimensioning).

I——X Finishing allowance (G12).

K——Z Finishing allowance (G12).

C——Cutting depth/per time.

F——Feed speed, mm/r.

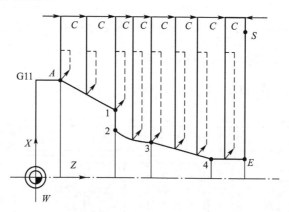

Figure 2.12　G11 Function

2. G11 Radial Volume and Surface-orientated Pre-finishing Machining

G11 X__(U__) Z__(W__) I__ K__ C__ (F__)

X——Start S: X (Absolute dimensioning).

U——Start S: X (Incremental dimensioning).

Z——Start S: Z (Absolute dimensioning).

W——Start S: Z (Incremental dimensioning).

I——X Finishing allowance (G12).

K——Z Finishing allowance (G12).

C——Cutting depth/per time.

F——Feed speed, mm/r.

3. G12 Finishing Machining

G12 X __ (U __) Z __ (W __) (F __)

X——Start S: X (Absolute dimensioning).
U——Start S: X (Incremental dimensioning).
Z——Start S: Z (Absolute dimensioning).
W——Start S: Z (Incremental dimensioning).

4. G13 Machining Cycle Call (Execution)

G13 N1= __ N2= __

N1——Definition contour, Program number: start.
N2——Definition contour, Program number: end.

5. G96 Constant Surface Speed Control

G97 Constant surface speed control Cancel
G96 S __ F __ D __

S——Linear velocity (m/min).
F——Feed speed.
D——Maximum speed (rpm).

G97 S __

S——Speed (rpm).

G10 Example (Figure 2.13):

```
% PM241001
N241001
N1 G54;                              (Definition of the programming zero point)
N2 G99 X140 Z127;                    (Blank definition)
N3 G96 S100 D2500 T1013 M4;          (Constant surface speed control, Linear veloci-
                                      ty 100m/min, Maximum speed 2500 rpm)
N4 G10 X145 Z130 I0.5 K0.5 C2.5 F0.5; (Definition, Finishing allowance X0.5  Z0.5, cut-
                                      ting depth/P2.5, Start 145, 130)
N8001 G01 X40 Z125;                  (Definition of contour Start)
N5 G01 W-5;
N6 G01 X30 W-5;
N7 G01 X50 Z95;
N8 G01 X30 Z75;
N9 G01 Z60;
N10 G01 X90;
N11 G01 U20 Z50;
N12 G01 U15;
N13 G01 X130 Z45;
N14 G01 Z35;
```

N15 G01 X140;
N16 G01 Z28;   (Definition of contour End)
N17 G13 N1= 8001 N2= 16;   (Program calls: Roughing the outside contour)
N18 G00 X200 Z145;   (Approaching the tool change position)
N19 T2023;   (Selection of the tools)
N20 G12 X145 Z127 S200;   (Finishing the outside contour: Definition)
N21 G13 N1= 8001 N2= 16;   (Program calls)
N22 G00 X300 Z350;   (Tool retracting)
N23 M30;

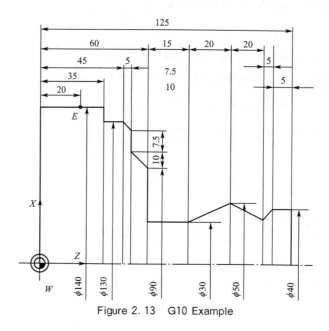

Figure 2.13 G10 Example

G11 Example (Figure 2.14):

% PM241002
N241002
N1 G54;   (Workpiece coordinate system)
N2 G99 X140 Z127;   (Blank definition)
N3 G96 S100 D2500 T1013 M4;   (Constant surface speed control, Linear velocity
                                    100m/min, Maximum speed 2500 rpm)
N4 G00 X200 Z145;   (Setting at the start point)
N5 G11 X145 Z120 I0.5 K0.5 C2.5 F0.5; (Definition, Finishing allowance X0.5;   Z0.5,
                                    Cutting depth/P2.5, Start 145, 120)
N6 G01 Z25;   (Definition of contour Start)
N7 G41 X140;   (Cutter compensation, Left)
N8 G01 Z35;
N9 G01 X130;
N10 G01 Z45 F0.05;
N11 G01 X125 Z50;
N12 G01 X110;

N13 G03 X90 Z60 R10;

N14 G01 Z75;

N15 G01 X63 Z95 F0.1;

N16 G01 X50;

N17 G01 Z115;

N18 G01 X40 Z120;   (Definition of contour End)

N19 G40;   (Cancel of compensation)

N20 G13 N1= 6 N2= 19;   (Program calls: Roughing the outside contour)

N21 G00 X200 Z145;   (Approaching the tool change position)

N22 T2023;   (Selection of the tools)

N23 G12 X145 Z120 S200;   (Finishing the outside contour: Definition)

N24 G13 N1= 6 N2= 19;   (Program calls)

N25 G00 X300 Z350;

N26 M30;

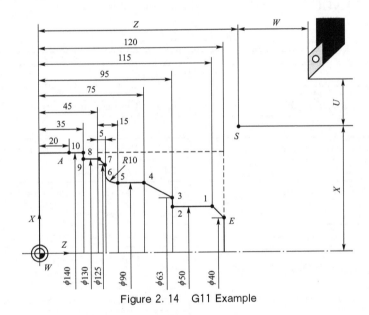

Figure 2.14 G11 Example

6. G32 Thread Cutting (Figure 2.15)

G32 X __(U __) Z __(W __) C __ (D __) (A __) (J __)(B __) F __

X——Bottom diameter of thread.

Z——End point of Z axis.

U——Depth of thread, +U internal thread, −U outside thread.

W——Length of thread.

C——1st cut depth, if C=U, Cut only one; If C<U, cut repetitious.

D——Cut depth of finishing.

A——Angle between threads/2.

J——End diameter of taper cut.

B——Taper, 1 : P.

F——Thread pitch.

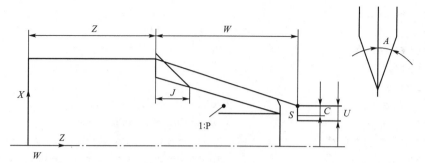

Figure 2.15 G32 Parameter

2.4.2 Turning Programming of FANUC 0i MATE TB

G codes of FANUC 0i MATE TB are shown in Table 2.4.

Table 2.4 G Codes

G code	Function	G code	Function
G00	Positioning (Rapid traverse)	G01	Linear interpolation (Cutting feed)
G02	Circular interpolation CW	G03	Circular interpolation CCW
G04	Dwell		
G10	Programmable data input	G11	Programmable data input mode cancel
G12	Input in inch	G13	Input in mm
G27	Reference position return check	G28	Return to reference position
G31	Skip function	G32	Thread cutting
G40	Tool nose radius compensation cancel	G41	Tool nose radius compensation left
G42	Tool nose radius compensation right	G50	Coordinate system setting or maximum spindle speed setting
G52	Local coordinate system setting	G53	Machine coordinate system setting
G54	Workpiece coordinate system 1 selection	G55	Workpiece coordinate system 2 selection
G56	Workpiece coordinate system 3 selection	G57	Workpiece coordinate system 4 selection
G58	Workpiece coordinate system 5 selection	G59	Workpiece coordinate system 6 selection
G65	Macro calling	G66	Macro modal call
G67	Macro modal call cancel	G70	Finishing cycle
G71	Stock removal in turning	G72	Stock removal in facing
G73	Pattern repeating	G74	End face peck drilling
G75	Outer diameter/internal diameter drilling	G76	Multiple threading cycle
G90	Outer diameter/internal diameter cutting cycle	G92	Thread cutting cycle
G94	End face turning cycle		
G96	Constant surface speed control	G97	Constant surface speed control cancel
G98	Per minute feed	G99	Per revolution feed

1. G01 Linear Interpolation (Cutting Feed)

G01 X(U)__ Z(W)__ F__

Example G01 (Figure 2.16):

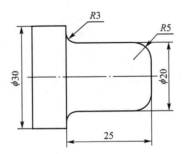

Figure 2.16 G01 Linear Interpolation

O2421
N10 T0101;
N20 G0 X0 Z1. S900 M03;   (Absolute command)
N30 G1Z0. F0.2;
N40 G1 X20. R-5. ;
N50 G1 Z-25. R3. ;
N60 G1 X30. 5;
N70 G28 X120. Z100. ;
N80 M30;

2. G04 Dwell

G04 X(U)__ or G04 P__

X—— Specify a time (decimal point permitted).

U ——Specify a time (decimal point permitted).

P ——Specify a time (decimal point not permitted).

3. G28 Return to Reference Position

G28 X(U)__ Z(W)__

Positioning to the intermediate or reference positions are performed at the rapid traverse rate of each axis. Therefore, for safety, the tool nose radius compensation, and tool offset should be cancelled before executing this command.

4. G32 Thread Cutting

G32 X(U)__ Z(W)__ F__

In general, thread cutting is repeated along the same tool path in rough cutting through finish cutting for a screw. Since thread cutting starts when the position coder mounted on the spindle outputs a one-turn signal, threading is started at a fixed point and the tool path on the workpiece is unchanged for repeated thread cutting. Note that the spindle speed must remain constant from rough cutting through finish cutting. If not, incorrect thread lead will occur.

Example G32 (Figure 2.17):

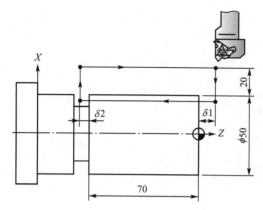

Lead of screw: 3mm
$\delta1$: 5mm
$\delta2$: 1.5mm
Depth of cut: 1mm(2cut two times)

Figure 2.17 G32

O2422
N10 G50;
N20 G97 S800 M03;
N30 G00 X90.0 Z5.0 T0202 M8;
N40 X48.0;
N50 G32 Z-71.5 F3.0;
N60 G00 X90.0;
N70 Z5.0;
N80 X46.0;
N90 G32 Z-71.5;
N100 G00 X90.0;
N110 Z5.0;
N120 X150.0 Z150.0;
N130 M30;

5. G50 Coordinate System Setting Spindle Speed Setting

Coordinate system setting or maximum spindle speed setting.

6. G90 Absolute and Incremental Programming (G code system B)

There are two ways to command travels of the tool: the absolute command and the incremental command. In the absolute command, coordinate value of the end position is programmed. In the incremental command, move distance of the position itself is programmed.

G90 and G91 are used to command absolute or incremental command, respectively.

G90 Outer diameter/internal diameter cutting cycle (G code system A)

G90 X(U)__ Z(W)__ F__

Example G90-1 (Figure 2.18):

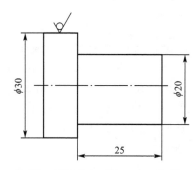

Figure 2.18 G90 Outer Diameter Cutting Cycle

O2423
N10 T0101;
N20 G0 X31. Z1. S800 M03;   (Start point)
N30 G90 X26. Z-24. 9 F0. 3;   (X 2mm, machining allowance 0. 1mm for finish machining)
N40 X22. ;
N50 X20. 5;   (X 0. 25mm)
N70 X20. Z-25. F0. 2 S1200;   (Finish machining)
N80 G28 X100. Z100. ;
N90 M30;

Example G90-2 (Figure 2.19):

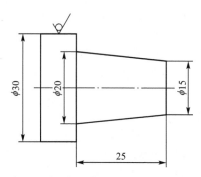

Figure 2.19 G90 Cone Processing

O2424
N10 T0101;   (Cutter)
N20 G0 X32. Z0. 5 S900 M3;
N30 G90 X26. Z-25. R-2. 5 F0. 15;   (Rough machining)
N40 X22. ;
N50 X20. 5;   (Machining allowance 0. 5mm for finish machining)
N60 G0 Z0 S1200 M3;
N70 G90 X20. Z-25. R-2. 5 F0. 1;
N80 G28 X100. Z100. ;
N90 M5;
N100 M2;

7. G92 Thread Cutting Cycle

G92 X(U)__ Z(W)__ F __

The range of thread leads, limitation of spindle speed and other parameters, are the same as in G32 (thread cutting). Thread chamfering can be performed in this thread cutting cycle. A signal from the machine tool initiates thread chamfering. The chamfering distance is specified in a range from 0.1L to 12.7L in 0.1L increments by parameter (No. 5130). (In the above expression, L is the thread lead).

Example G92-1 (Figure 2.20):

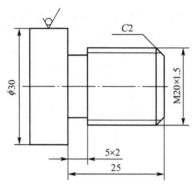

Figure 2.20　G92 Thread Cutting Cycle

```
O2425
N110 T0303;
N120 G0 X28. Z5. S900 M3;       (Start point)
N130 G92 X19.4 Z-23. F1.5;      (Thread cutting)
N140 X19.;                      (Cycle)
N150 X18.6;
N160 X18.2;
N170 X18.;
N180 X17.9;
N190 X17.8;
...
```

Example G92-2 (P=1.5) (Figure 2.21):

G92 X(U)__ Z(W)__ R __ F __

Lead (L) is specified.

R —— Rapid traverse.

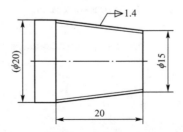

Figure 2.21　G92 Taper Screw Thread Process

O2426
N10 T0101;
N20 G0 X25. Z5. S900 M3;
N30 G92 X19. 6 Z-20. R-2. 5 F1. 5;
N40 X19. 4;
N50 X19. ;

8. G94 End Face Turning Cycle

G94 X(U)__ Z(W)__ F__

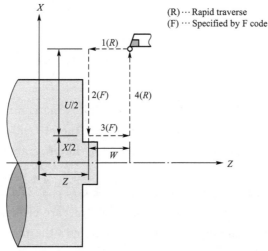

Figure 2. 22 Parameter of End Face Turning Cycle

In incremental programming, the sign of numbers following addresses U and W depends on the direction of paths 1 and 2. That is, if the direction of the path is in the negative direction of the Z axis, the value of W is negative.

In single block mode, operations 1, 2, 3 and 4 are performed by pressing the cycle start button once.

G94 can be used as taper face cutting cycle.

Example G94 (Figure 2.23):

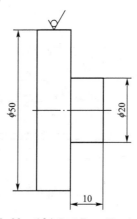

Figure 2. 23 G94 End Face Turning Cycle

```
O2427
N10 T0101;
N20 G0 X52. Z1. S900 M03;
N30 G94 X20.2 Z-2. F0.2;    (Rough machining, Z 2)
N40 Z-4.;
N50 Z-6.;
N60 Z-8.;
N70 Z-9.8;
N80 X20. Z-10. S1200;       (Finish machining)
N90 G28 X100. Z100.;
N100 M30;
```

9. G70 Finishing Cycle

G70 P(ns)__ Q(nf)__

After rough cutting by G71, G72 or G73, the following command permits finishing.

(ns) —— Sequence number of the first block for the program of finishing shape.

(nf) —— Sequence number of the last block for the program of finishing shape.

NOTE:

(1) F, S and T functions specified in the block G71, G72, G73 are not effective but those specified between sequence numbers "ns" and "nf" are effective in G70.

(2) When the cycle machining by G70 is terminated, the tool is returned to the start point and the next block is read.

(3) In blocks between "ns" and "nf" referred in G70 through G73, the subprogram cannot be called.

10. G71 Stock Removal in Turning

G71 U(Δd)__ R(e)__
G71 P(ns)__ Q(nf)__ U(Δu)__ W(Δw)__ F__ S__ T__

If a finished shape of A to A' to B is given by a program as in the figure below, the specified area is removed by Δd (depth of cut), with finishing allowance Δu/2 and Δw left (Figure 2.24).

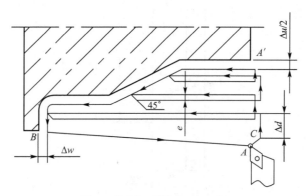

Figure 2.24 G71 Parameter

Δd——Depth of cut (radius designation).

Designate without sign. The cutting direction depends on the direction AA'. This designation is modal and is not changed until the other value is designated. Also this value can be specified by the parameter (No. 5132), and the parameter is changed by the program command.

e——Escaping amount.

This designation is modal and is not changed until the other value is designated. Also this value can be specified by the parameter (No. 5133), and the parameter is changed by the program command.

ns——Sequence number of the first block for the program of finishing shape.

nf——Sequence number of the last block for the program of finishing shape.

Δu——Distance and direction of finishing allowance in X direction (diameter/radius designation).

Δw——Distance and direction of finishing allowance in Z direction.

F, S, T——Any F, S or T function contained in blocks "ns" to "nf" in the cycle is ignored, and the F, S or T function in this G71 block is effective.

Example G71 (Figure 2.25):

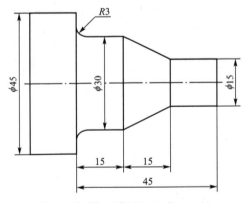

Figure 2.25 G71 Stock Removal

```
O2428
N10 T0101;
N20 G0 X46. Z0.5 S900 M03;
N30 G71 U2. R0.5;                    (Depth of cut 2mm)
N40 G71 P50 Q110 U0.3 W0.1 F0.3;    (Machining allowance of finish machining X 0.3mm,
                                     Z 0.1mm; Feed speed of rough machining 0.3mm/r)
N50 G1 X15.;
N60 G1 Z0. F0.15 S1500;              (Feed speed of finish machining 0.15mm/r; 800rpm)
N70 Z-15.;
N80 X30. Z-30.;
N90 Z-42.;
N100 G2 X36. Z-45. R3.;
```

N110 G1 X46.;
N120 G70 P50 Q100;   (Finish machining cycle)
N130 G28 X100. Z100.;
N140 M5;
N150 M30;

11. G72 Stock Removal in Facing

G72 W(d)__ R(e)__
G72 P(ns)__ Q(nf)__ U(u)__ W(w)__ F__ S__ T__

As shown in Figure 2.26, this cycle is the same as G71 except that cutting is made by an operation parallel to X axis.

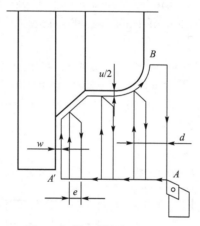

Figure 2.26 G72 Parameter

Example G72 (Figure 2.27):

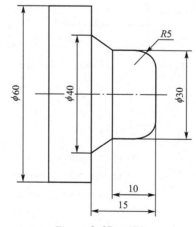

Figure 2.27 G72

(Diameter designation, metric input)

O2429
N10 T0101;

N20 G0 X61. Z0. 5 S900 M03;
N30 G72 W2. R0. 5;
N40 G72 P50 Q100 U0. 1 W0. 3 F0. 25;
N50 G0 Z-15. ;
N60 G1 X40. F0. 15 S1500;
N70 X30. Z-10. ;
N80 Z-5. ;
N90 G2 X20. Z0 R5. ;
N100 G0 Z0. 5;
N110 G70 P60 Q110;
N120 G28 X100. Z100. ;
N130 M30;

12. G73 Pattern Repeating

G73 U(Δi)__ W(Δk)__ R(Δd)__
G73 P(ns)__ Q(nf)__ U(Δu)__ W(Δw)__ F__ S__ T__

This function permits cutting a fixed pattern repeatedly, with a pattern being displaced bit by bit. By this cutting cycle, it is possible to efficiently cut work whose rough shape has already been made by a rough machining, forging or casting method, and so on.

Example G73 (Figure 2. 28):

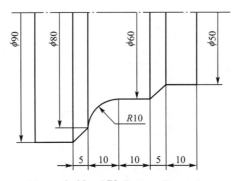

Figure 2. 28   G73 Pattern Repeating

O24210
N10 T0101;
N20 G0 X110. Z10. S800 M3;
N30 G73 U5. W3. R3. ;
N40 G73 P50 Q110 U0. 4 W0. 1 F0. 3;
N50 G0 X50. Z1. S1000;
N60 G1 Z-10. F0. 15;
N70 X60. Z-15. ;
N80 Z-25. ;
N90 G2 X80. Z-35. R10. ;
N100 G1 X90. Z-40. ;
N110 G0 X110. Z10. ;

N120 G70 P50 Q110;
N130 G28 X100. Z150. ;
N140 M30;

13. G74 End Face Peck Drilling

G74 R(e)__
G74 X(u)__ Z(w)__ P(Δi)__ Q(Δk)__ R(Δd)__ F __

e——Return amount.

This designation is modal and is not changed until the other value is designated. Also this value can be specified by the parameter No. 5139, and the parameter is changed by the program command.

X——X component of point B.

U——Incremental amount from A to B.

Z——Z component of point C.

W——Increment amount from A to C.

Δi——Movement amount in X direction (without sign), P1000 is 1mm.

Δk——Depth of cut in Z direction (without sign).

Δd——Relief amount of the tool at the cutting bottom. The sign of Δd is always plus (+).

However, if address X(U) and Δi are omitted, the relief direction can be specified by the desired sign.

F——Feed rate.

Example G74:

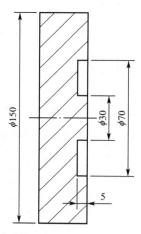

Figure 2.29 G74 End Face Peck Drilling

O2430
N10 T0606;   (Width of cutter 4mm)
N20 S900 M3;
N30 G0 X30. Z2. ;
N40 G74 R1. ;
N50 G74 X62. Z-5. P3500 Q3000 F0. 1;

N60 G0 X200. Z50. M5;
N70 M30;

Example (Figure 2.30):

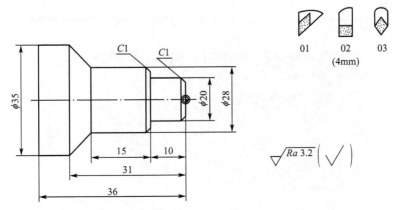

Figure 2.30 Application of G Code

```
O2431                           (O Program number)
N10 T0101;
N20 S1000 M03;
N30 G00 X40. Z2. ;
N40 G71 U1.5 R1. ;              (U-Depth of cut, R-Escaping amount)
N50 G71 P60. Q150. U0.5 W0.2 F0.3;   (P-Sequence number of the first block, Q- Sequence
                                      number of the last block, U-X machining allowance,
                                      W-Z machining allowance)
N60 G01 X18. Z0. ;              (Start point)
N80     X20. Z-1. ;             (Chamfer 1×45°)
N90 G01     Z-10. ;             (φ20)
N100    X26. ;
N110    X28. Z-11. ;            (Chamfer 1×45°)
N120        Z-25. ;
N140    X35. Z-31. ;
N150        Z-36. ;             (End point)
N160 G70 P60 Q150 S1500 F0.05;  (N60—N150)
N170 G00 X50. Z60. ;
N180 T0202;
N190 S200 M03;                  (Cut off by right edge)
N210 G00 X37. Z-40. ;           (Length 36mm)
N220 G01 X0. F0.03;
N230 G00 X50. ;
N240 G00 Z0. ;
N250 M05;
N260 M02;                       (End)
```

2.4.3 Milling Programming of MAHO 600C

MAHO 600C have four axis ($B\ X\ Y\ Z$). The interface is shown in Figure 2.31. G codes of MAHO 600C are shown in Table 2.5.

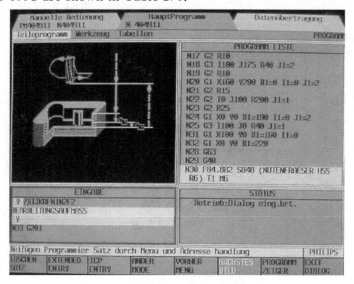

Figure 2.31 MAHO 600C

Table 2.5 G codes of MAHO 600C

G code		Meaning
G00	*	Rapid travel
G01		Linear interpolation
G02		Clockwise circular interpolation CW
G03		Counter-clockwise circular interpolation CCW
G04	* *	Dwell 0, 1-983 Sec
G11	* *	Contour element (round corner) sequence programming
G14	* *	Repeat section selection: Unconditional (single or multiple) or conditional program section repetition
G17	*	Plane selection XOY
G18		Plane selection XOZ
G19		Plane selection YOZ
G22	* *	Subprogram selection
G23	* *	Main program selection
G25		Manual feed correction (F-over) ON
G26		Manual feed correction (F-over) OFF
G27		Feed with rounding
G28		Feed with precision stop

（续）

G code		Meaning
G40	*	Cancel of G41, G42, G43, G44
G41		Tool radius compensation left
G42		Tool radius compensation right
G43		Tool radius compensation up to
G44		Tool radius compensation beyond
G51	*	Cancel of G52
G52		Selection of the zero point shift G52 stored at setting of workpiece zero point (Reset Axis)
G53	*	Cancel for G54, G55, G56, G57, G58, G59
G54		Selection of stored Zero point shift G54
G55		Selection of stored Zero point shift G55
G56		Selection of stored Zero point shift G56
G57		Selection of stored Zero point shift G57
G58		Selection of stored Zero point shift G58
G59		Selection of stored Zero point shift G59
G63	*	Cancel of G64
G64		Activation of the geometry processor
G70		Dimensions in inches (Inch dimension system)
G71	*	Dimensions in mm (mm dimension system)
G72	*	Cancel of G73
G73		Mirror image machining, scale up and scale down (In combination with the word A4=)
G74	* *	Travel to an absolute position referred directly to the reference point R of the machine in rapid traverse
G77	* *	Cycle selection with point circle definition
G78	* *	Point definition
G79	* *	Cycle selection
G81		Drilling cycle
G83		Drilling cycle with chip clearance/chip breaking (Deep hole drilling cycle)
G84		Tapping cycle
G85		Reaming cycle
G86		Boring cycle
G87		Rectangular recess milling cycle
G88		Slot milling cycle
G89		Circular recess milling cycle
G90	*	Absolute dimensions (Reference dimensions)
G91		Incremental dimensions (Chain dimensions)

(续)

G code		Meaning
G92		Programmed incremental zero point shift. Rotation of the coordinate system (In conjunction with the word B4=)
G93		Programmed absolute zero point shift. Rotation of the coordinate system (In conjunction with the word B4=)
G94	*	Feed speed in mm per minute (G71) or in inches per minute (G70)
G95		Rotational feed in mm per revolution (G71) or in inches per revolution (G70)
G98		Definition of window for graphic test runs
G99		Definition of workpiece for graphic test runs

* Functions are active in the read condition of the CNC control system.

** Active only in sentence.

Example: Milling Thickness 20 (Figure 2.32).

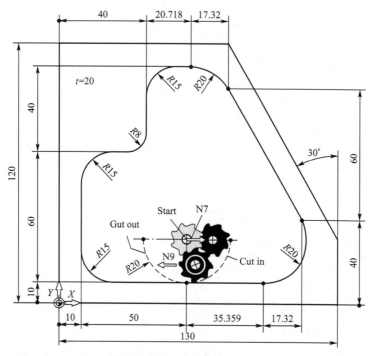

Figure 2.32 Milling example

```
% PM24301
N24301
N1 G17 S900 T31 M66;              (Tool diameter 10mm)
N2 G54;                           (Selection of stored zero point shift  G54)
N3 G98 X-10 Y-10 Z-20 I150 J140 K30;   (Display window)
N4 G99 X0 Y0 Z-20 I130 J120 K20;       (Workpiece definition)
N5 G00 X60 Y30 Z8 M3;
N6 G01 Z-21 F50;                  (Down)
```

N7 G43 X80 F100;	(Tool radius compensation up to X80)
N8 G42;	(Tool the compensation right)
N9 G02 X60 Y10 R20;	(Or I60 J30)
N10 G01 X25;	
N11 G02 X10 Y25 R15;	(Or I25 J25)
N12 G01 Y55;	
N13 G02 X25 Y70 R15;	(Or I25 J55)
N14 G01 X32;	
N15 G03 X40 Y78 R8;	(Or I32 J78)
N16 G01 Y95;	
N17 G02 X55 Y110 R15;	(Or I55 J95)
N18 G01 X60.718;	
N19 G02 X78.039 Y100 R20;	(Or I60.718 J90)
N20 G01 X112.679 Y40;	
N21 G02 X95.359 Y10 R20;	(Or I90.359 J30)
N22 G01 X60;	
N23 G02 X40 Y30 R20;	(Or I60 J30)
N24 G40;	(Cancel the compensation)
N25 G00 Z50 M5;	(Approaching the tool change position)
N26 G53;	(Cancel of G54)
N27 T0 M6;	(Manual tool change)
N28 M30;	

1. G40/G41/G42 Cutter Compensation

In CNC machining, if the cutter axis is moving along the programmed path, the dimension of the workpiece obtained will be incorrect since the diameter of the cutter has not be taken into account.

Modern CNC systems are capable of doing this type of calculation which is known as cutter compensation. What the system requires are the programmed path, the cutter diameter and the position of the cutter with reference to the contour. Normally, the cutter diameter is not included in the program. It has to be input to the CNC system in the tool setting process.

If the cutter is on the left of the contour, G41 is used. The tool is positioned on the left hand side of the part, as seen following the direction of movement, from behind the tool.

If the cutter is on the right of the contour, G42 will be used. The tool is positioned on the right hand side of the part, as seen following the direction of movement, from behind the tool.

G40 is to cancel the compensation calculation.

2. G14 Repeat Section Selection: Unconditional (single or multiple) or Conditional Program Section Repetition

G 14 N1= __ N2= __ J __

N1——Program number: start.

N2——Program number: end.

J——Repeat times (default once).

3. G92, G93 Zero Point Shift (Figure 2.33)

G92 X＿ Y＿ Z＿ B1=＿ L1=＿ B4=＿

Programmed incremental zero point shift. Rotation of the coordinate system (In conjunction with the word B4=)

G93 X＿ Y＿ Z＿ B2=＿ L2=＿ B4=＿

Programmed absolute zero point shift. Rotation of the coordinate system (In conjunction with the word B4=)

X, Y, Z——Displacement.

B1, L1——Polar (G92).

B2, L2——Polar (G93).

B4——The angle between two axis.

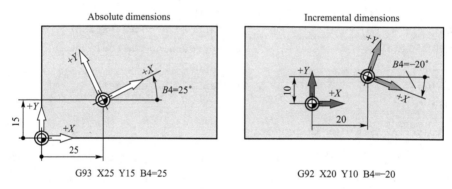

Figure 2.33 Shift of Coordinate System

Example: Milling Thickness 20 (Figure 2.34).

```
% PM24302
N24302
N1 G17 S800 T31 M66;              (Tool diameter 10mm)
N2 G54;
N3 G98 X-10 Y-10 Z-10 I260 J215 K30;
N4 G99 X0 Y0 Z-20 I240 J195 K20;  (Workpiece definition)
N5 G00 X55 Y45 Z2 M13;
N6 G01 Z-21 F50;                  (Down)
N7 G43 Y55 F100;                  (Tool radius compensation up to Y55)
N8 G42;                           (Tool compensation right)
N9 G01 X97;                       (Milling the contour)
N10 G02 X105 Y47 R8;
N11 G01 Y18;
N12 G02 X97 Y10 R8;
N13 G01 X83;
```

```
N14 G02 X75 Y18 R8;
N15 G01 Y30;
N16 G01 X35;
N17 G01 Y18;
N18 G02 X27 Y10 R8;
N19 G01 X23;
N20 G02 X15 Y18 R8;
N21 G01 Y47;
N22 G02 X23 Y55 R8;
N23 G01 X55;                      (Milling the contour End)
N24 G00 Z50;
N25 G40;                          (Cancel compensation)
N26 G92 Y55;                      (Programmed incremental zero point shift)
N27 G14 N1= 5 N2= 26 J2;          (Repeat twice)
N28 G93 X120 Y130;                (Programmed absolute zero point shift)
N29 G14 N1= 5 N2= 25;             (Repeat once)
N30 G93 X145 Y10 B4= 30;          (Programmed absolute zero point shift)
N31 G14 N1= 5 N2= 25;             (Repeat once)
N32 G00 Z50 M5;                   (Approaching the tool change position)
N33 G53;                          (Cancel of G54)
N34 M30;
```

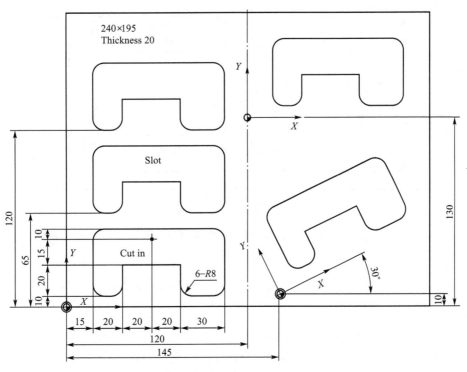

Figure 2.34 Shift Example

4. G72, G73 Mirror Image Machining, Scale up and Scale down

G72   Cancel of G73
G73 X-1 (Y-1) (Z-1)   Mirror image machining (Figure 2.35)
G73 A4= __   Scale up and scale down (In combination with the word A4=)

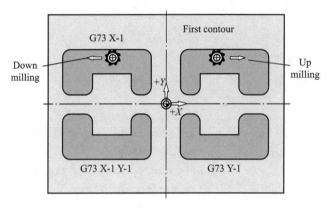

Figure 2.35　Mirror Image

Example mirror (Figure 2.36):

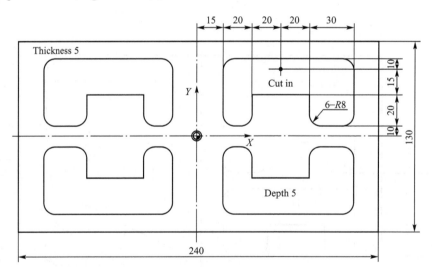

Figure 2.36　Mirror Image Example

```
% PM24303
N24303                              (Workpiece 240×130×5)
N1 G17 S800 T31 M66;                (Tool diameter 10mm)
N2 G54;
N3 G98 X-130 Y-75 Z-10 I260 J150 K20;
N4 G99 X-120 Y-65 Z-5 I240 J130 K5;  (Workpiece definition)
N5 G0 X55 Y45 Z2 M13;               (G00)
N6 G1 Z-6 F50;                      (Down)
N7 G43 Y55 F100;                    (Tool radius compensation up to Y55)
```

N8 G42;	(Tool compensation right)	
N9 G1 X97;	(Milling the contour)	
N10 G2 X105 Y47 R8;		
N11 G1 Y18;		
N12 G2 X97 Y10 R8;		
N13 G1 X83;		
N14 G2 X75 Y18 R8;		
N15 G1 Y30;		
N16 G1 X35;		
N17 G1 Y18;		
N18 G2 X27 Y10 R8;		
N19 G1 X23;		
N20 G2 X15 Y18 R8;		
N21 G1 Y47;		
N22 G2 X23 Y55 R8;		
N23 G1 X55;	(Milling the contour End)	
N24 G0 Z50;		
N25 G40;	(Cancel compensation)	
N26 G73 X-1;	(Mirror Y)	
N27 G14 N1= 5 N2= 25;	(Repeat once)	
N28 G72;		
N29 G73 X-1 Y-1;	(Mirror Origin)	
N30 G14 N1= 5 N2= 25;	(Repeat once)	
N31 G72;		
N32 G73 Y-1;	(Mirror X)	
N33 G14 N1= 5 N2= 25;	(Repeat once)	
N34 G72;	(Cancel of G73)	
N35 G0 Z50 M5;	(Approaching the tool change position)	
N36 G53;		
N37 M30;		

2.4.4 Milling Programming of FANUC 0MC

G codes of FANUC 0MC are shown in Table 2.6.

Table 2.6 G Codes of FANUC 0MC

G code	Function	G code	Function
G00	Positioning	G01	Linear interpolation
G02	Circular interpolation/Helical interpolation CW	G03	Circular interpolation/Helical interpolation CCW
G04	Dwell, Exact stop	G09	Exact stop
G10	Programmable data input	G11	Programmable data input mode cancel
G15	Polar coordinates command cancel	G16	Polar coordinates command

(续)

G code	Function	G code	Function
G17	XOY plane selection	G18	XOZ plane selection
G19	YOZ plane selection		
G20	Input in inch	G21	Input in mm
G22	Stored stroke check function on	G23	Stored stroke check function off
G24	Mirror image	G25	Mirror image cancel
G27	Reference position return check	G28	Return to reference position
G29	Return from reference position	G30	2nd, 3rd and 4th reference position return
G31	Skip function	G33	Thread cutting
G39	Corner offset circular interpolation		
G40	Cutter compensation cancel/Three dimensional compensation cancel	G41	Cutter compensation left/Three dimensional compensation
G42	Cutter compensation right	G43	Tool length compensation + direction
G44	Tool length compensation − direction	G49	Tool length compensation cancel
G50	Scaling cancel	G51	Scaling
G52	Local coordinate system setting	G53	Machine coordinate system selection
G54	Workpiece coordinate system 1 selection	G55	Workpiece coordinate system 2 selection
G56	Workpiece coordinate system 3 selection	G57	Workpiece coordinate system 4 selection
G58	Workpiece coordinate system 5 selection	G59	Workpiece coordinate system 6 selection
G65	Macro call	G66	Macro modal call
G67	Macro modal call cancel		
G68	Coordinate rotation/Three dimensional coordinate conversion	G69	Coordinate rotation cancel/Three dimensional coordinate conversion cancel
G73	Peck drilling cycle	G74	Counter tapping cycle
G76	Fine boring cycle		
G80	Canned cycle cancel/External operation function cancel	G81	Drilling cycle, spot boring cycle or external operation function
G82	Drilling cycle or counter boring cycle	G83	Peck drilling cycle
G84	Tapping cycle	G85	Boring cycle (Rough)
G86	Boring cycle	G87	Back boring cycle
G90	Absolute command	G91	Increment command
G92	Setting for work coordinate system or clamp at maximum spindle speed		
G94	Feed per minute	G95	Feed per rotation
G96	Constant surface speed control	G97	Constant surface speed control cancel
G98	Return to initial point in canned cycle		

1. G90/G91 Absolute and Incremental Programming

There are two ways to command travels of the tool, the absolute command, and the incremental command. In the absolute command, coordinate value of the end position is programmed. In the incremental command, move distance of the position itself is programmed. G90 and G91 are used to command absolute or incremental command, respectively.

2. G00 Positioning (Rapid Traverse)

G00 X__ Y__ Z__

3. G01 Linear Interpolation

G01 X__ Y__ Z__ F__

4. G02/G03 Circular Interpolation or Helical Interpolation

G02 X__ Y__ R__ F__
G02 X__ Y__ I__ J__ F__

The end point of an arc is specified by address X, Y or Z, and is expressed as an absolute or incremental value according to G90 or G91. For the incremental value, the distance of the end point which is viewed from the start point of the arc is specified.

The arc center is specified by addresses I, J and K for the X, Y and Z axes, respectively.

5. G04 Dwell

G04 X__ or G04 P__

(Dwell with the time in seconds or speed specified)
For dwell with the speed specified, another option is required.
G04 X3.5 or G04 P3500, Dwell 3.5s.

6. Cutter Compensation

G40 G01 X__ Y__
G41(G42)G01 X__ Y__ D__

D—— Code for specifying as the cutter compensation value (1-3digits) (D code).
G41 Cutter compensation left;
G42 Cutter compensation right;
G40 Cutter compensation cancel.

When the tool is moved, the tool path can be shifted by the radius of the tool.

To make an offset as large as the radius of the tool, CNC first creates an offset vector with a length equal to the radius of the tool (start-up). The offset vector is perpendicular to the tool path. The tail of the vector is on the workpiece side and the head positions to the center of the tool.

7. G92 Setting for Work Coordinate System or Clamp at Maximum Spindle Speed

G92 X__ Y__ Z__

A workpiece coordinate system is set by specifying a value after G92 in the program. Example (Figure 2.37):

G92 X20. Y10. Z10. ;

8. G53 Machine Coordinate System Selection

G53 G90 X __ Y __ Z __

When a command is specified the position on a machine coordinate system, the tool moves to the position by rapid traverse. G53, which is used to select a machine coordinate system, is a one-shot G code; that is, it is valid only in the block in which it is specified on a machine coordinate system.

Example (Figure 2.38):

G53 G90 X-100. Y-100. Z-20. ;

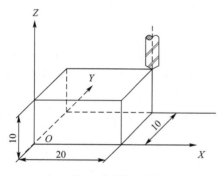

Figure 2.37　G92

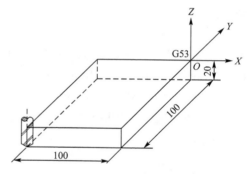

Figure 2.38　G53

9. G54, G55, G56, G57, G58, G59 Workpiece Coordinate System 1-6 Selection

G54 G90 G00 (G01) X __ Y __ Z __ (F __)

Workpiece coordinate system 1 to 6 are established after reference position return and the power is turned on. When the power is turned on, G54 coordinate system is selected.

Choosing from six workpiece coordinate systems set using the CRT/MDI panel.

Example (Figure 2.39):

G54: X-50. Y-50. Z-10. ;
G55: X-100. Y-100. Z-20. ;
Origin O'-G54, Origin O''-G55.

N10 G53 G90   X0. Y0. Z0. ;
N20 G54 G90 G01 X50. Y0. Z0. F100;
N30 G55 G90 G01 X100. Y0. Z0. F100;   (Trajectory of tool nose —OAB)

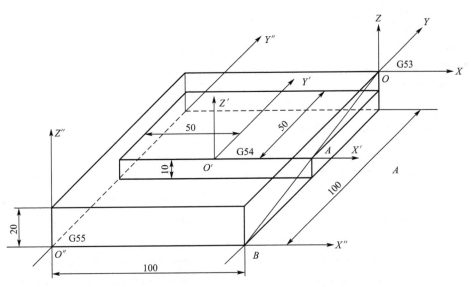

Figure 2.39 Set Workpiece Coordinate System

Example of workpiece coordinate system (Figure 2.40):
Tool radius: 5mm, depth of cut: 5mm.

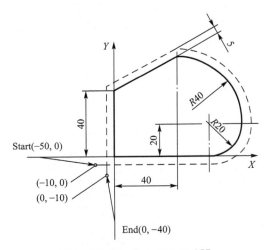

Figure 2.40 Example of G55

```
Set G55: X=-400., Y=-150., Z=-50.; H01=5.
O24401
N10 G55 G90 G01 Z40. F200;   (Workpiece coordinate system 2 selection)
N20 M03 S900;
N30 G01 X-50. Y0.;           (Start of X, Y)
N40 G01 Z-5. F100;           (Start of Z)
N50 G01 G42 X-10. Y0. H01;   (Cutter compensation right)
N60 G01 X60. Y0.;            (Cut in)
N70 G03 X80. Y20. R20.;
N80 G03 X40. Y60. R40.;
```

```
N90 G01 X0. Y40. ;
N100 G01 X0. Y-10;            (Cut out)
N110 G01 G40 X0. Y-40. ;      (Cutter compensation cancel)
N120 G01 Z40. F200;
N130 M05;
N140 M30;                     (End)
```

10. G68, G69 Coordinate Rotation

G68 X __ Y __ R __
G69 coordinate rotation cancel.

X, Y——Absolute command for two of the X, Y and Z axes that correspond to the current plane selected by a command (G17, G18 or G19). The command specifies the coordinates of the center of rotation for the values.

R——Angular displacement with a positive value indicates counter clockwise rotation.

A programmed shape can be rotated. By using this function it becomes possible, for example, to modify a program using a rotation command when a workpiece has been placed with some angle rotated from the programmed position on the machine. Furthermore, when there is a pattern comprising some identical shapes in the positions rotated from a shape, the time required for programming and the length of the program can be reduced by preparing a subprogram of the shape and calling it after rotation.

Example of coordinate rotation (Figure 2.41):

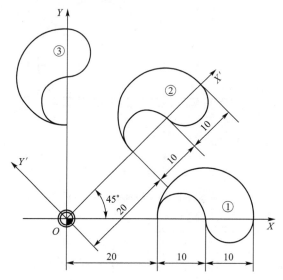

Figure 2.41 Coordinate Rotation

```
O24402                        (Main program)
N10 G90 G17 S900 M03;
N20 M98 P1100;                (①)
N30 G68 X0. Y0. P45;          (Coordinate rotation 45°)
N40 M98 P1100;                (②)
```

N50 G69;   (Coordinate rotation cancel)
N60 G68 X0. Y0. P90;   (Coordinate rotation 90°)
M70 M98 P1100;   (③)
N80 G69 M05 M30;   (Coordinate rotation cancel)

Subprogram (①)
O1100
N100 G90 G01 X20. Y0. F100;
N110 G02 X30. Y0. R5. ;
N120 G03 X40. Y0. R5. ;
N130 X20. Y0. R10. ;
N140 G00 X0. Y0. ;
N150 M99;

11. M98 Subprogram Call

M98 P __

P——Number of times the subprogram is called repeatedly.
When no repetition data is specified, the subprogram is called just once.
M99 Program end.

12. G51 Scaling

G51 X __ Y __ Z __ P __
G50

X, Y, Z——Absolute command for center coordinate value of scaling.
P——Scaling magnification: 0.001—999.999.

13. G24/G25 Programmable Mirror Image

Example of mirror image (Figure 2.42):

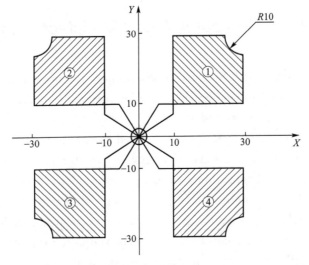

Figure 2.42   Mirror Image

```
O24403                          (Main program)
N10 G91 G17 S900 M03;
N20 M98 P1101;                  (①)
N30 G51.1 X0.;                  (Y axis, X=0)
N40 M98 P1101;                  (②)
N50 G51.1 X0. Y0.;              (X axis, Y axis, (0, 0))
N60 M98 P1101;                  (③)
N70 G50.1 X0.;                  (Cancel a programmable mirror image Y)
N80 G51.1 Y0.;                  (X axis, Y0)
N90 M98 P1101;                  (④)
N100 G50.1 Y0.;                 (Canceling a programmable mirror image)
N110 M05;
N120 M30;

Subprogram (①):
%1101
N200 G41 G00 X10.0 Y4.0 D01;
N210 Y1.0;
N220 Z-98.0;
N230 G01 Z-7.0 F100;
N240 Y25.0;
N250 X10.0;
N260 G03 X10.0 Y-10.0 I10.0;
N270 G01 Y-10.0;
N280 X-25.0;
N290 G00 Z105.0;
N300 G40 X-5.0 Y-10.0;
N310 M99;
```

2.5 Computer Aided Manufacturing

In manual preparation of a CNC part programming, the programmer is required to define the machine or the tool movement in numerical terms. If the geometry is complicated 3D surfaces cannot be programmed manually.

Over the past years, lots of efforts are devoted to automate the part program generation. With the development of the CAD/CAM system, interactive graphic system is integrated with the CNC part programming. Graphic based software using menu driven technique improves the user friendliness. The part programmer can create the geometrical model in the CAM package or directly extract the geometrical model from the CAD/CAM data base. Built-in tool motion commands can assist the part programmer to calculate the tool paths automatically. The part programmer can verify the tool paths through the graphic display using the animation function of the CAM system. It greatly enhances the speed and accuracy in tool path generation.

The task flow that is needed for producing a part using an NC machine can be summarized as shown in Figure 2.43. The tasks can be classified as the following three types:

(1) Offline tasks: CAD, CAPP, CAM.
(2) Online tasks: CNC machining, monitoring and On-machine measurement.
(3) Post-line tasks: Computer-Aided Inspection (CAI), post-operation.

Offline tasks are the tasks that are needed to generate a part program for controlling a CNC machine. In the offline stage, after the shape of a part has been decided, a geometry model of this part is created by 2D or 3D CAD. In general, CAD means Computer Aided Design, but CAD in this book is regarded as a modeling stage in which both design and analysis are included because engineering analysis of a part cannot be carried out on the shop floor.

After finishing geometric modeling, Computer Aided Process Planning (CAPP), is carried out where necessary information for machining is generated. In this stage, the selection of machine tools, tools, jig and fixture, decisions about cutting conditions, scheduling and machining sequences are created. Because process planning is very complicated and CAPP is immature with respect to technology, process planning generally depends on the know-how of a process planner.

CAM (Computer Aided Manufacturing) is executed in the final stage for generating a part program. In this stage, tool paths are generated based on geometry information from CAD and machining information from CAPP. During tool path generation, interferences between tool and workpiece, minimization of machining time and tool change, and machine performance are considered. In particular, CAM is an essential tool to generate 2.5D or 3D tool paths for machine tools with more than three axes.

Online tasks are those that are needed to machine parts using CNC machines. A part program, being the machine-understandable instructions, can be generated in the above-mentioned offline stage and part program for a simple part can be directly edited in CNC by the user. In this stage, the CNC system reads and interprets part programs from memory and controls the movement of axes. The CNC system generates instructions for position and velocity control based on the part programs and servo motors are controlled based on the instructions generated. As the rotation of a servo motor is transformed into linear movement via ball-screw mechanisms, the workpiece or tool is moved and, finally, the part is machined by these movements.

To increase the machining accuracy, not only the accuracy of the servo motor, table guide, ball screw and spindle but also the rigidity of the machine construction should be high. The construction of the machine and the machine components should also be designed to be insensitive to vibration and temperature. In addition, the performance of the encoder and sensors that are included in the CNC system and the control mechanism influences the machining accuracy. The control mechanism will be addressed in more detail in the following section.

In the online stage, the status of the machine and machining process may be monitored during machining. Actually, tool-breakage detection, compensation of thermal deformation, adaptive control, and compensation of tool deflection based on monitoring of cutting force, heat, and electric current are applied during machining. On-machine measurement is also used to calculate machining error by inspecting the finished part on the machine, returning machining errors to CNC to carry out compensation.

The post-line task is to carry out CAI, inspecting the finished part. In this stage, inspection using a CMM (Coordinate Measurement Machine) is used to make a comparison between the result and the geometry model in order to perform compensation. The compensation is executed by modifying tool compensation or by doing post-operations such as re-machining and grinding. Reverse engineering, meaning that the shape of the part is measured and a geometric model based on the measured data is generated, is included in this stage. As mentioned above, through three stages, it is possible for machine tools not only to satisfy high accuracy and productivity but also to machine parts with complex shape as well as simple shapes. Because CNC machines can machine a variety of parts by changing the part program and repetitively machine the same part shape by storing part program, CNC machines can be used for general purposes.

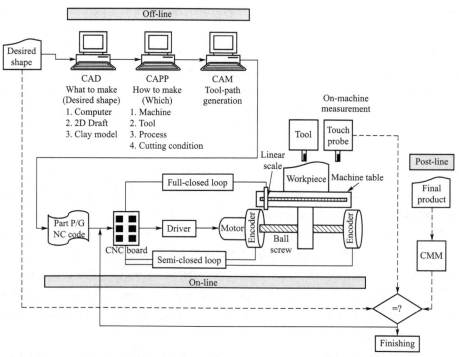

Figure 2.43　The Architecture of CNC Machine Tools and Machining Operation Flow

There are several computer-aided manufacturing or CAD/CAM system available in the market. Their basic features can be summarized below:

(1) Geometric modeling/CAD interface.

(2) Tool motion definition.

(3) Data processing.

(4) Post processing.

(5) Data transmission.

The flow chart of a CAM system is shown in Figure 2.44.

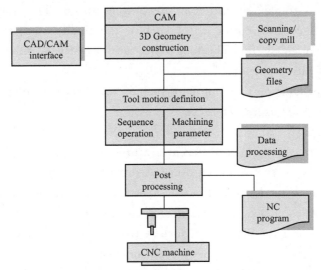

Figure 2.44   Flow Chart of a CAM System

1) Geometric modeling/CAD interface

The geometry of the workpiece can be defined by basic geometrical elements such as points, lines, arcs, splines or surfaces. The two dimensional or three dimensional geometrical elements are stored in the computer memory in forms of a mathematical model. The mathematical model can be a wire frame model, a surface model, or a solid model.

In addition, the geometric models can be imported from other CAD/CAM system through standard CAD/CAM interface formats such as Initial Graphic Exchange Specification (IGES). IGES is a graphic exchange standard jointly developed by industry and the National Bureau of Standards with the support of the U.S. Air Force. It provides transportability of 3-dimensional geometry data between different systems.

Throughout this system, geometrical elements from one system can be translated into a neutral file standard and then from this standard into other format.

2) Tool motion definition

After the geometric modeling, machining data such as the job setup, operation setup and motion definition are input into the computer to produce the cutting location file (CL file) for machining the workpiece.

(1) Job setup: It is to input the machine datum, home position, and the cutter diameters for the CL file.

(2) Operation setup: It is to input into the system the operation parameters such as the feed rate, tolerance, and approach/retract planes, spindle speed, coolant ON/OFF, stock offset and the tool selection etc.

(3) Motion definition: Built-in machining commands are used to control the tool mo-

tion to machine the products. This includes the whole processing, profile machining, pocketing, surface machining, gouge checking, etc.

3) Data processing

The input data is translated into computer usable format. The computer will process the desired part surface, the cutter offset surface and finally compute the paths of the cutter which is known as CL file. The tool paths can normally be animated graphically on the display for verification purpose.

Furthermore, production planning data such as tool list, set up sheet, and machining time is also calculated for users' reference.

4) Post processing

Different CNC machines have different features and capabilities, the format of the CNC program may also vary from each other. A process is required to change the general instructions from the cutter location file to a specific format for a particular machine tool and this process is called post processing. Post processor is a computer software which converts the cutter location data files into a format that the machine controller can interpret correctly. Generally, there are two types of post processor.

(1) Specific post processor. This is a tailor-made software which outputs the precise code for a specific CNC machine. The user is not required to change anything in the program.

(2) Generic (universal) post processor.

This is a set of generalized rules which needs the user to customize into the format that satisfies the requirements of a specific CNC machine.

5) Data transmission

After post processing, the CNC program can be transmitted to the CNC machines either through the off-line or on-line process.

(1) Off-line processes. Data carriers are used to transmit the CNC program to the CNC machines. It includes paper tapes, magnetic tape or magnetic disc.

(2) On-line processes. On-line processes is commonly used in DNC operation and data is transferred either serially or parallel using data cables.

6) Serial transmission

Asynchronous serial transmission is most widely used in data transmission and RS-232C is the most popular asynchronous standard. Built-in RS-232C serial port (9 pins or 25 pins) is available in many computers. RS-232C is inexpensive, easy to program and with a baud rate up to 38400. However, its noise margin is limited up to 15 meters.

Parallel transmission is commonly used in data transmission between computers and external devices such as sensors, programmable logic controllers (PLC) or actuators. One common standard is IEEE488. It contains a 24 lines bus with 8 for data, 8 for controls and 8 for ground. It can transfer data up to 1 Mbps for a 20-meter cable.

Local Area Networks (LAN)

To enable the CAD/CAM facilities to run smoothly, it is desirable for the facilities to be linked together. In the local-area network, terminals can access any computer on the network or devices on the shop floor without a physical hardwire with speed up to 300 megabits per second. For instance, Ethernet runs at 100Mbps which is much faster than an RS-232 serial communication (115. 2kbps).

A LAN consists of both software and hardware design, which are governed by a set of rules called protocol. The design of software enables control of data to be handled and error to be recovered while hardware generates and receives signals, and media carries the signals. The protocol defines the logical, electrical, and physical specifications of the network. The same protocol must be followed in order to have an effective communication with each other in the network.

Exercises

1. Read and match each of the following pictures to the name of the hand tool.

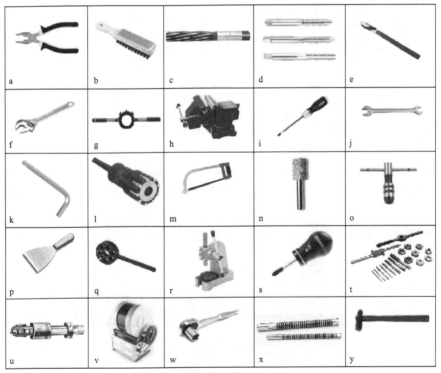

Figure 2.45　Exercise 1

[ ] hacksaw   [ ] bench vise   [ ] file card   [ ] scraper   [ ] tap & die
[ ] hand taps   [ ] hammer   [ ] burnishing tool   [ ] adjustable wrench [ ] tap extractor
[ ] hex key   [ ] arbor press   [ ] T-handle tap wrench [ ] hand reamer   [ ] screwdriver
[ ] pliers   [ ] external lap   [ ] rotary files   [ ] Phillips   [ ] open-end
                                                                   screwdriver   wrench
[ ] broaches   [ ] socket wrench [ ] tumbling machine   [ ] die stock   [ ] file

2. Learn the structure of vernier calipers (Figure 2.46) and give the reads of Figure 2.47.

Vernier calipers are precision tools used to make accurate measurements to within 0.001 inch or 0.02 mm. They can measure the outside diameter or width of an object. They can measure the inside diameter or width of an object. They can also measure the depth of an object. Vernier calipers are available in inch and metric graduations. Some types have both inch and metric graduations on the same caliper. The vernier caliper consists of one L-shaped frame and one movable jaw. The L-shaped frame consists of a bar and the fixed jaw. The bar shows the main scale graduations. The movable jaw slides along the bar and contains the vernier scale. An adjusting nut adjusts for size and the locking or clamp screws lock readings into place.

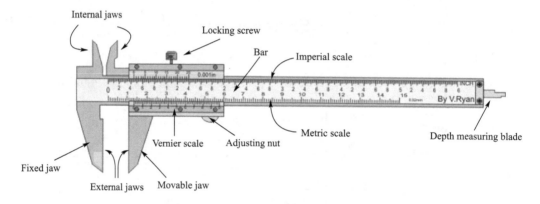

Figure 2.46 Exercise 2(1)

Most bars are graduated on both sides or on both edges. One side takes outside measurements and the other takes inside measurements. The internal jaws measure inside spaces like holes. The external jaws measure things like the outside diameter of a piece of pipe.

Look at the two figures below and determine the readings.

Figure 2.47 Exercise 2(2)

This reads _____

3. Look at the following micrometer and fill in the blanks.

Sleeve reads full mm =_____ .
Sleeve reads 1/2 mm =_____ .
Thimble reads =_____ .

CNC Part Programming Chapter 2

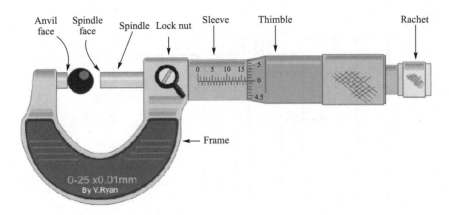

Figure 2.48 Exercise 3(1)

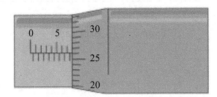

Figure 2.49 Exercise 3(2)

Total measurement = _____ .

4. Narrate CNC coordinate systems.
5. Give the CNC program of following Figure 2.50 (units: mm).

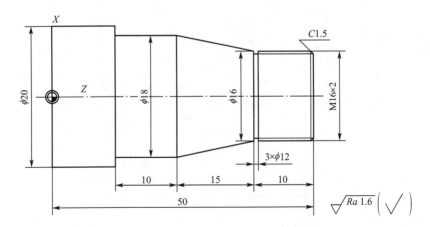

Figure 2.50 Exercise 5

6. Give the CNC program of following Figure 2.51 (units: mm).
7. Give the CNC program of following Figure 2.52 (units: mm).

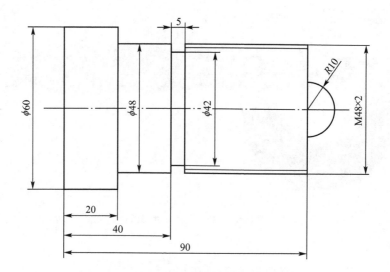

Figure 2.51　Exercise 6

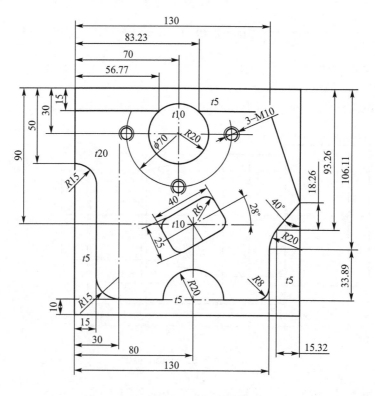

Figure 2.52　Exercise 7

8. Give the CNC program of following Figure 2.53 (units: mm).

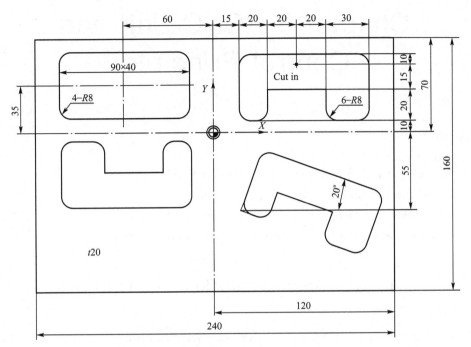

Figure 2.53 Exercise 8

Chapter 3 CNC Unit and Control Principle

Objectives

- To understand the hardware structure of a CNC unit.
- To understand the concept of interpolation.
- To master the concept of cutter radius compensation.
- To understand the CNC acceleration-deceleration control.

3.1 Hardware Architecture of a CNC Unit

CNC unit is the core of CNC system. The hardware comprising a CNC unit is made up of microprocessors, electronic memory modules, I/O interfaces, and position control modules, etc., which is just like an ordinary computer system. A CNC unit reaches machine through control position transducers, tool holding device, work holding device and machine tool body (Figure 3.1).

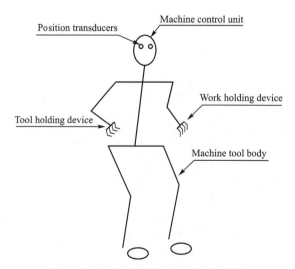

Figure 3.1 A CNC Unit and its Device

Figure 3.2 shows the data flow between the modules. The data is transmitted between the modules via the ring buffer defined by global variables, and the ring buffers are located between Interpreter and Rough Interpolator, between Rough Interpolator and

Acceleration/Deceleration (abbreviation: Acc/Dec) Controller, and between Acc/Dec Controller and Fine Interpolator. Each ring buffer includes the data is shown in Figure 3.2. The Fine Interpolator and Position Controller use global variables to send the necessary data.

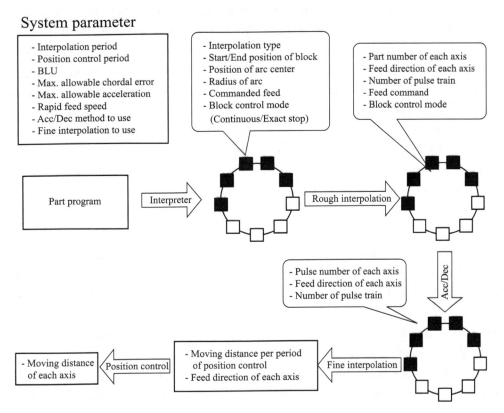

Figure 3.2   Data Flow between Modules

Features of CNC:

(1) Storage of more than one part program: With improvements in computer storage technology, newer CNC controllers have sufficient capacity to store multiple programs. Controller manufacturers generally offer one or more memory expansions as options to MCU.

(2) Various forms of program input: Whereas conventional (hard-wired) MCU are limited to punched tape as the input medium for entering part program. CNC controllers generally possess multiple data entry capabilities, such as punched tape, magnetic tape, floppy diskettes, RS-232 communications with external computers, and manual data input (operator entry of program).

(3) Program editing at the machine tool: CNC permits a part program to be edited while it resides in the MCU computer memory. Hence, a part program can be tested and corrected entirely at the machine site, rather than being returned to the programming office for corrections. In addition to part program corrections, editing also permits cutting conditions in the machining cycle to be optimized. After the program has been corrected and

optimized, the revised version can be stored on punched tape or other media for future use.

(4) Fixed cycles and programming subroutines: The increased memory capacity and the ability to program the control computer provide an opportunity to store frequently used machining cycles such as macros that can be called by the part program. Instead of writing the full instructions for the particular cycle into every program, a programmer includes a call statement in the part program to indicate that the macro cycle should be executed. These cycles often require that certain parameters be defined, for example, a bolt hole circle, in which the diameter of the bolt circle, the spacing of the bolt holes, and other parameters must be specified.

(5) Interpolation: Some of the interpolation schemes are normally executed only on a CNC system because of computational requirements. Linear and circular interpolations are sometimes hard-wired into the control unit, but helical, parabolic, and cubic interpolations are usually executed by a stored program algorithm.

(6) Positioning features for setup: Setting up the machine tool for a given work part involves installing and aligning a fixture on the machine tool table. This must be accomplished so that the machine axes are established with respect to the work part.

The alignment task can be facilitated using certain features made possible by software options in the CNC system. Position set is one of the features. With position set, the operator is not required to locate the fixture on the machine table with extreme accuracy. Instead, the machine tool axes are referenced to the location of the fixture using a target point or set of target points on the work or fixture.

(7) Cutter length and size compensation: In older style controls, cutter dimensions have to be set precisely to agree with the tool path defined in the part program. Alternative methods for ensuring accurate tool path definition have been incorporated into the CNC controls. One method involves manually entering the actual tool dimensions into the MCU. These actual dimensions may differ from those originally programmed. Compensations are then automatically made in the computed tool path. Another method involves use of a tool length sensor built into the machine. In this technique, the cutter is mounted in the spindle and the sensor measures its length. This measured value is then used to correct the programmed tool path.

(8) Acceleration and deceleration calculations: This feature is applicable when the cutter moves at high feed rates. It is designed to avoid tool marks on the work surface that would be generated due to machine tool dynamics when the cutter path changes abruptly. Instead, the feed rate is smoothly decelerated in anticipation of a tool path change and then accelerated back up to the programmed feed rate after the direction change.

(9) Communications interface: With the trend toward interfacing and networking in plants today, most modern CNC controllers are equipped with a standard RS-232 or other communications interface to link the machine to other computers and computer driven devices. This is useful for various applications, such as:

① Downloading part program from a central data file.

② Collecting operational data such as workpiece counts, cycle times, and machine utilization.

③ Interfacing with peripheral equipment, e. g. robots that unload and load parts.

(10) Diagnostics: Many modern CNC systems possess a diagnostics capability that monitors certain aspects of the machine tool to detect malfunctions or signs of impending malfunctions or to diagnose system breakdowns.

3.1.1 MCU for CNC

The MCU is the hardware that distinguishes CNC from conventional NC. The general configuration of the MCU in a CNC system is illustrated in Figure 3.3. The MCU consists of the following components and subsystems:

(1) Central processing unit.
(2) Memory.
(3) Input/Output interface.
(4) Controls for machine tool axes and spindle speed.
(5) Sequence controls for other machine tool functions.

These subsystems are interconnected by means of a system bus, which communicates data and signals among the components of network. The machine control unit is the heart of a CNC system. There are two sub-units in the machine control unit: the Data Processing Unit (DPU) and the Control Loop Unit (CLU).

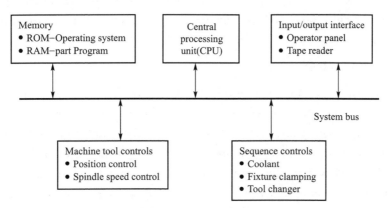

Figure 3.3　Configuration of CNC Machine Control Unit

1. Central Processing Unit

The central processing unit (CPU) is the brain of the MCU. It manages other components in the MCU based on software contained in main memory. The CPU can be divided into three sections: ①control section, ②arithmetic-logic unit, and ③immediate access memory.

The control section retrieves commands and data from memory and generates signals to activate other components in the MCU. In short, it sequences, coordinates, and regulates all the activities of the MCU computer. The arithmetic-logic unit (ALU) consists of the circuitry to perform various calculations (addition, subtraction and multiplication), counting, and logical functions required by software residing in memory. The immediate access memory provides a temporary storage of data being processed by the CPU. It is connected to main memory of the system data bus.

2. Memory

The immediate access memory in the CPU is not intended for storing CNC software. A much greater storage capacity is required for various programs and data needed to operate the CNC system. As with most other computer systems, CNC memory can be divided into two categories: ① main memory, and ② secondary memory. Main memory (also known as primary storage) consists of ROM and RAM devices. Operating system software and machine interface programs are generally stored in ROM. These programs are usually installed by the manufacturer of the MCU. Numerical control part programs are stored in RAM devices. Current programs in RAM can be erased and replaced by new programs as jobs are changed.

High-capacity secondary memory (also called Auxiliary Storage or Secondary Storage) devices are used to store large programs and data files, which are transferred to main memory as needed. Common among the secondary memory devices are hard disks and portable devices that have replaced most of the punched paper tape traditionally used to store part programs. Hard disks are high-capacity storage devices that are permanently installed in the CNC machine control unit. CNC secondary memory is used to store part programs, macros, and other software.

3. Input/Output (I/O) Interface

The I/O interface provides communication software between the various components of the CNC system, other computer systems, and the machine operator. As its name suggests, The I/O interface transmits to and receives data and signals from external devices. The operator control panel is the basic interface by which the machine operator communicates to the CNC system. This is used to enter commands related to part program editing, MCU operating mode (e.g., program control vs. manual control), speeds and feeds, cutting fluid pump on/off, and similar functions. Either an alphanumeric keypad or keyboard is usually included in the operator control panel. The I/O interface also includes a display (CRT or LED) for communication of data and information from the MCU to the machine operator. The display is used to indicate current status of the program as it is being executed and to warn the operator of any malfunctions in the CNC system.

As indicated previously, CNC part programs are stored in a variety of ways. Programs can also be entered manually by the machine operator or stored at a central computer site and transmitted via LAN to the CNC system. Whichever means is employed by the plant, a suitable device must be included in the I/O interface to allow input of the program into MCU memory.

4. Controls for Machine Tool Axes and Spindle Speed

These are hardware components that control the position and velocity (feed rate) of each machine axis as well as the rotational speed of the machine tool spindle. The control signals generated by MCU must be converted to a form and power level suited to the particular position control systems used to drive the machine axes. Positioning systems can be classified as open loop or closed loop, and different hardware components are required in each case.

Depending on the type of machine tool, the spindle is used to drive either ① workpiece or ② a rotating cutter. Turning exemplifies the first case, whereas milling and drilling exemplify the second. Spindle speed is a programmed parameter for most CNC machine tools. Spindle speed components in the MCU usually consist of drive control circuit and a feedback sensor interface. The particular hardware components depend on the type of spindle drive.

5. Sequence Controls for Other Machine Tool Functions

In addition to control of table position, feed rate, and spindle speed, several additional functions are accomplished under part program control. These auxiliary functions are generally on/off (binary) actuations, interlocks, and discrete numerical data. To avoid overloading the CPU, PLC is sometimes used to manage the I/O interface for these auxiliary functions.

3.1.2 New Technology for CNC Control

1. GPU

A graphics processing unit (GPU), also occasionally called Visual Processing Unit (VPU), is a specialized electronic circuit designed to rapidly manipulate and alter memory to accelerate the creation of images in a frame buffer intended for output to a display. GPU are used in embedded systems, mobile phones, personal computers, workstations, and game consoles. Modern GPU are very efficient at manipulating computer graphics, and their highly parallel structure makes them more effective than general-purpose CPU for algorithms where processing of large blocks of data is done in parallel. In a personal computer, a GPU can be present on a video card, or it can be on the motherboard or in certain CPU on the CPU die.

The term GPU was popularized by NVIDIA Corporation in 1999, who marketed the GeForce 256 as "the world's first 'GPU', or Graphics Processing Unit, a single-chip processor with integrated transform, lighting, triangle setup/clipping, and rendering engines that are capable of processing a minimum of 10 million polygons per second". Rival ATI Technologies coined the term visual processing unit or VPU with the release of the RADEON 9700 in 2002.

Modern GPU use most of their transistors to do calculations related to 3D computer graphics. They were initially used to accelerate the memory-intensive work of texture mapping and rendering polygons, later adding units to accelerate geometric calculations such as the rotation and translation of vertices into different coordinate systems. Recent developments in GPU include support for programmable shaders which can manipulate vertices and textures with many of the same operations supported by CPU, over sampling and interpolation techniques to reduce aliasing, and very high-precision color spaces. Because most of these computations involve matrix and vector operations, engineers and scientists have increasingly studied the use of GPU for non-graphical calculations. An example of GPU being used non-graphically is the generation of Bitcoins, where the graphical processing unit is used to solve hash functions.

In addition to the 3D hardware, today's GPU include basic 2D acceleration and frame buffer capabilities (usually with a VGA compatibility mode). Newer cards like AMD/ATI HD5000-HD7000 even lack 2D acceleration. It has to be emulated by 3D hardware.

2. Motion Controllers

A motion controller controls the motion of some object. Frequently motion controllers are implemented using digital computers, but motion controllers can also be implemented with only analog components as well.

The role of a motion controller is to control the high-speed and high-precision motion of a linear motor in accordance with instructions from a CNC unit. It holds the industrial property rights to copy and use the Proportion Integration Differential (PID) calculus control that realizes high-speed and high-acceleration control of a linear motor, the technology for manufacturing a motion controller for a linear motor using modern control theories and software for controlling a linear motor.

In addition to the development of CNC units, which dictate the overall performance of a machine, it is of great importance for enhancing machining precision to give and transmit instructions as accurate as possible to the muscles (linear motors). What is needed to calculate the motion of a linear motor accurately and transmit the calculation results to the CNC units is a motion controller. The development of the motion controller started aiming at working out tailored units designed specifically for machine tool instead of modifying existing ones.

3. Multi-axis Motion Control Technology

In a machine with multiple axes (Figure 3.4), the motions of individual axes must often be coordinated. A simple example would be a robot that needs to move two joints to reach a new position. We could extend the motion of the slower joints so that the motion of each joint would begin and end together.

Motion controllers require a load (something to be moved), a prime mover (something to cause the load to move), some sensors (to be able to sense the motion and monitor the prime mover), and a controller to provide the intelligence to cause the prime mover to move the load as desired. Motion controllers are used to achieve some desired ben-

Figure 3.4 Multi-axis Motion Controller

efit which can include:
(1) Increased position and speed accuracy.
(2) Higher speed.
(3) Faster reaction time.
(4) Increased production.
(5) Smoother movement.
(6) Reduction in cost.
(7) Integration with other automation.
(8) Integration with other processes.
(9) Ability to convert desired specifications into motion required to produce a product.
(10) Increased information and ability diagnose and troubleshoot.
(11) Increased consistency.
(12) Improved efficiency.
(13) Elimination of hazards to humans or animals.

3.2 CNC System Software

A CNC machine consists of two major components: the machine tool and the controller (or the CNC unit), which is an on-board computer. These computers may be or may not be manufactured by the same company. General Numeric, Fanuc, General Electric, Bendix, Cincinnati Milacron, and Siemens are among those manufacturers of CNC controllers that supply units to makers of machine tools. Each controller is manufactured with a standard set of building codes, i.e., start-up program, basic system codes, machining and measuring loop programs, etc. Other codes are added by the machine tool builders. For this reason, program codes vary somewhat from machine to machine. Every CNC machine, regardless of manufacturer, is a collection of systems coordinated by the controller.

The main functions of a CNC system are mainly brought out by control software. Control software does the following work by corresponding subroutines.

(1) Compiling the part codes input by users.

(2) Interpolation calculations.

(3) Tool compensation.

(4) Speed control.

(5) Position control.

Software in a CNC system is shown in Figure 3.5.

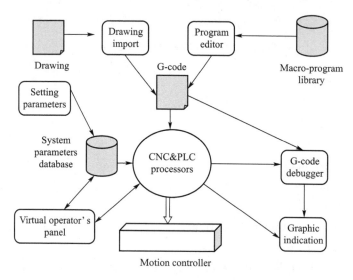

Figure 3.5 Software in a CNC System

In a CNC system, there are various kinds of software codes, which fall into two groups: user software and system software.

1. User Software

User software is also called part program, which is programmed and input by using numerical control language (such as APT) to indicate the machining procedure of the part. It is compiled into a program with various G-codes and M-codes arranged according to the process sequence.

2. System Software

The main functions of a CNC system are mainly brought out by control software. Control software does the following work by corresponding subroutines.

1) Compiling the part codes input by users

After receiving the part codes, which are written in ISO or EIA format, compiler will translate, trim and store them into a specified format, decode the machining instructions in part codes and do decimal-binary transitions to coordinate data, calculate tool's center path considering the offset of tool radius, pre-calculate some constants that will be used in interpolation calculation and speed control process, etc. Of course, the contents of pre-process are varied in different CNC systems, but they have the same objectives of saving time for real-time interpolation calculations. It is not very exigent for CNC systems to pre-process the input part codes in real time. It can be done before machining or in the idle intervals while machining. The more sufficiently when processing inputted part codes, the more smoothly when doing real time calculations of interpolation and speed control.

2) Interpolation calculations

Interpolation calculation subroutine has the same functions with hardware interpolator in NC system, i.e., to assign electrical pulses for axes. It is a strict real time program which demands as less as possible instruction codes, to shorten the time for performing interpolation calculations, which is because that the time is determinative to the feed speed of interpolation. In some CNC system, it is adopted to combine rough interpolation and fine interpolation, i.e., software is used for rough interpolation that interpolates a tiny line each time. And then hardware does fine interpolation by turning the tiny line into a series of single pulses and outputting the pulses. By this means, the feed speed can be improved, and the computer from complicated interpolation calculation to other necessary process can be liberated.

3) Tool compensation

All types of CNC machine tools require some form of compensation, such as tool length compensation, cutter radius compensation, and tool nose radius compensation. Though applied for different reasons on different machine types, all forms of compensation allow the CNC user to modify in unpredictable conditions related to tooling. Generally speaking, if the CNC user is faced with any unpredictable situations during programming, it is likely that the CNC control manufacturer has come up with a form of compensation to deal with the problem.

4) Speed control

Speed control subroutine aims at controlling the speed of pulses assigning, i.e., controlling the frequency of interpolation calculation according to the setting speed code (or corresponding speed instruction) to guarantee the preset feed velocity. While there is an unexpected abrupt changing in velocity during machining, the speed control subroutine should automatically speed up or slow down the velocity to avoid pace-loss in the drive system.

Speed control can be implemented totally by software method (software timer method), and can also be implemented by hardware means, i.e., using velocity code controlling an oscillator, then by means of interrupts or queries, the CNC unit does once interpolation calculation to guarantee the feed velocity. Furthermore, by using software processing the speed control data, and combining with the speed integrator hardware, the high performance of constant compound velocity controlling can be realized, and the feed velocity can be improved greatly.

5) Position control

Position control is in the position loop of the servo system. This work can be done by either software or hardware. Position control software compares the interpolation-calculated position with the real measured position in each sampling cycle, and controls the motor using the difference. Furthermore, position control software can usually adjust the magnification of the position loop circuit, compensate the error of pitch of screws in each direction of axes, and the non-return-to-zero when reversing the motion. Thus, the location precision can be improved.

3.3 Interpolation

Every CNC machine discussed in this book possesses more than one axis of motion. More often than not, it will be necessary for programmer to command that two or more axes move simultaneously in a controlled manner. For example, an end mill cutter is used on a machining straight surfaces, angular surfaces and round surfaces. While some of the movements in this example may involve only one axis, the angular and circular motions must involve at least two axes.

There are many enterprises that produce CNC systems in the world, in which Siemens (German), Fanuc (Japan), and Allen-Bradley (America, A-B) company have great influence on China's manufacturing automation. China also has its own brands of CNC that are widely used for manufacturing and education purposes, such as HNC (Huazhong Numerical Control System), Blue-Sky (Shenyang Institute of Computing Technology), etc.

Three familiar types of CNC units which are widely used in China are Fanuc, Siemens, and HNC.

In the early day of NC, the program was required to produce an angular or circular surface, the motion had to be broken down into a long series of very small one-axis motions to form the angle or circle as closely as possible to the desired shape. This kind of motion normally required the help of a computer to produce. With the advent of a feature called motion interpolation, programming common complex movements become much simpler. With today's current CNC controls, it is relatively easy to command angular and circular motion.

What is interpolation? The one that most applies to CNC is the math-related definition, "To estimate a missed functional values at neighboring points." When the control interpolates a motion, it is precisely estimating the programmed path base on a small amount of input data.

The Control Principle of Numerical Machine Tools

Linear interpolation consists of any programmed points linked together by straight lines, whether the points are close together or far apart. Curves can be produced with linear interpolation by breaking them into short, straight-line segments. This method has limitations, because a very large number of points would have to be programmed to describe the curve in order to produce a contour shape.

1. Point by Point Comparison Method Interpolation

When the control makes a straight motion in two axes, all that is required is the start point and the end point of the motion. The control automatically and instantaneously fills in the missing point between the start point and end point. What really happens is that the control makes a series of very small one-axis movement from the start point to the end point. This series of motions resembles a stairway. Each step along the way is very small, and the end result will appear to be a straight line. As you can see in Figure 3.6, when two or more axes are programmed, the control forms a series of small one-axis movement. The series of each step determines the resolution of the axis. The smaller the step, the better the resolution.

2. Point by Point Comparison Method Circular Arcs Interpolation

The stairs approximation algorithm, termed an incremental interpolator, determines the direction of the step every interval and sends the pulse to the related axis. In this section, the stairs approximation interpolator for a circle will be addressed and the algorithm for a line can be easily determined from the algorithm for a circle. Figure 3.7 shows how the stairs approximation interpolator for a circle behaves in the case that the commanded circular movement is in the clockwise direction in the first quadrant with respect to the center of the circle.

$$F_m = X_m^2 + Y_m^2 - R^2 \tag{3.1}$$

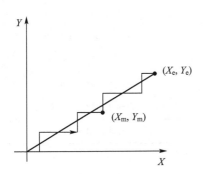
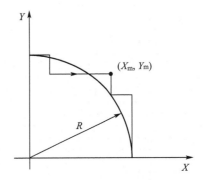

Figure 3.6 Point by Point Comparison Method Interpolation

Figure 3.7 The Behavior of the Stairs Approximation Interpolation Algorithm

The direction of a step is determined based on F_m, the commanded circular direction, and the quadrant where movement is done. For example, if the circular movements carried out in a clockwise direction in the first quadrant, the algorithm executed is as below.

$F_m > 0$, Step $-Y$; This case means that the position (X_m, Y_m) is located on the outside of a circle. In this case, the step is moved in the negative direction of the Y axis.

$F_m = 0$, Step $+X$; One of the above rules can be arbitrarily selected and applied.

$F_m < 0$, Step $+X$; This case means that the position (X_m, Y_m) is located on the inside of a circle. In this case, the step moves in the positive direction of X axis.

After one step is completed by applying the above rules, the position (X_{m+1}, Y_{m+1}) is updated and the procedure repeated until the tool reaches the commanded position (X_f, Y_f).

3.4 Forms of Compensation

Now we begin our discussions of the specific types of compensation available for different CNC machine tools. If you are a beginner of CNC, it is necessary that you understand the reasons why the various forms of compensation work as they do. In fact, it is important to understand why you need compensation for the various purposes. If you understand why you need the compensation type, you are well on your way to understanding how to use it. Also, knowing why compensation is used for its various purposes will allow you to easily adapt to any version of its use.

Table 3.1 is a list of the compensation types of the cutter radius compensation and the CNC machine tools related to the compensation.

Table 3.1 The Compensation Types

Compensation	Related
Tool length compensation	Machining centers
Cutter radius compensation	Machining centers
Fixture offsets	Machining centers
Dimensional tool offsets	Turning centers
Tool nose radius compensation	Turning centers
Wire radius compensation	Wire EDM machines
Wire taper compensation	Wire EDM machines

3.4.1 Tool Compensation

Using tool compensation values is easy to program a work part without consideration of the actually applicable tool lengths or tool radii. The available work part drawing data can be directly used for programming.

The tool data, lengths as well as radii of the milling machines are automatically considered by the CNC control.

1. Tool Length Compensation for Milling and Turning

A tool length compensation regarding the reference point enables the adjustment between the set and actual tool length. This tool length value has to be available for the control. It is necessary to measure the length L, i. e. the distance between the tool setup point B and the cutting tip, and to enter it into the control.

In case of milling tools, the length is defined in Z direction (Figure 3.8).

B——Tool setup point.

L——Length = distance of the cutting tip to the tool setup point in Z.

R——Radius of the milling tool.

In case of lathe tools, the length L is defined in Z direction (Figure 3.9).

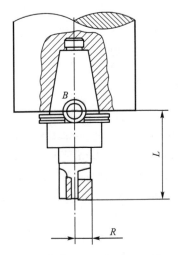

Figure 3.8 Tool Compensation Values on a Milling Tool

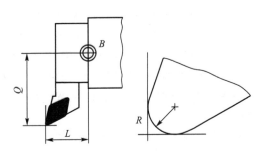

Figure 3.9 Tool Compensation Values on a Lathe Tool

B——Tool setup point.

L——Length = distance of the cutting tip to the tool set-in point in Z.

Q——Overhang = distance of the cutting tip to the tool setup point in X.

R——Cutting radius.

In the CNC control these tool compensation values are stored in the compensation value storage, whereby in most CNC controls, it is possible to describe up to 99 tools. These values have to be activated during machining.

This is done by calling the data within the CNC program, e. g., with the address H or by specific places in the T word.

2. Cutter Radius Compensation

Cutter radius compensation (also called Cutter Diameter Compensation) is used on machining centers and similar CNC machines. This feature allows the programmer to forget about the cutting tool's radius or diameter during programming. Like all forms of compensation, it makes programming easier, since the programmer doesn't have to concern about the exact cutter diameter while the program is being prepared. Cutter radius compensation also allows the radius of the cutting tool to vary without modification to the program (Figure 3.10).

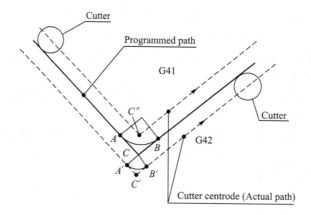

Figure 3.10 Cutter Radius Compensation

How does the cutter radius compensation work?

Understanding how the CNC control interprets cutter radius compensation commands will be the first step in solving any cutter radius compensation problem. Though there are some minor differences related to how each control manufacturer internally handles cutter radius compensation, the basic points we make in this section will apply to most current CNC controls.

Using cutter radius compensation involves three basic programming steps:

(1) Instate cutter radius compensation.

(2) Make movements to machine workpiece.

(3) Cancel cutter radius compensation.

Cutter radius compensation is instated with a command that tells the control how to position the cutter relative to the surfaces being machined throughout its movements. Either the cutter will be positioned to the left of the surface (with a G41) or to the right of the surface (with a G42). The man can easily remember G41 and G42 if he knows the difference between climb milling and conventional milling. If using a right hand cutter (spindle rotating clockwise with M03), climb milling is instated with G41 and conventional milling is instated with G42. Once instated, the control will keep the cutter to the left or right side of a series of lines and circles generated with straight line (G01) and circular (G02 and G03) commands. These lines and circles represent the actual surfaces being machined.

Cutter radius compensation will remain in effect until cancelled. That is, the cutter will be kept on the left side or right side of all motion commands until this cancellation. The command to cancel cutter radius compensation is G40.

What is limitation of cutter radius compensation?

Cutter radius compensation is not applied to all forms of cutting tools. It is needed only for cutting tools that have the ability to machine on the periphery of the cutter, and only when machining the periphery of the cutter. Tools like end mill, shell mills, and some face mills have this ability. Drills, reamers, taps, boring bars, and other center cutting tools have absolutely no use for cutter radius compensation.

There are many times during milling operations when the programmer wishes to machine the edge of a work-piece. This edge can be in the form of a straight surface or in the form of a contoured surface. Like a woodworking router, a milling cutter is driven along the edge of the workpiece.

One way to program the milling cutter's path is to program the motions by the centerline of the milling cutter. In this case, the programmer must take into consideration the diameter of the milling cutter. For example, if the milling cutter is 1 inch in diameter, all motions programmed must be kept precisely 0.5 inch away from the surfaces to be milled. Even this assumes that there is no tool pressure pushing the cutting tool away from its programmed path.

While using the centerline coordinates of the tool path is a popular way of programming, it has several disadvantages and limitation.

Here we list and explain them. Each of these limitations presented real problems before cutter radius compensation became available tool's centerline coordinates and part coordinates. Notice that the tool's centerline coordinates require at least one extra calculation to be made for each axis in the coordinate system. On the other hand, part coordinates are usually taken right from the print.

3.4.2 Reasons for Cutter Radius Compensation

There are several reasons why cutter radius compensation is so helpful. It relieves several programming burdens. Here we list and explain why it is so important to use this feature when you have reason to do so.

Reasons for cutter radius compensation:

(1) Part programming is easy.

(2) Roughing program is related to finish machining.

(3) Error compensation of the cutter.

Change of tool diameter. Using a 1-inch diameter end mill to machine the right side of a rectangular work-piece being held in a vise would be considered a simple operation by most experienced programmers. If a programmer prepares the program on the basis of the tool's centerline coordinates, not using cutter radius compensation, the end mill must be kept away from the right side of the rectangular work-piece by precisely 0.5 inch throughout its motions.

Imagine that the operator is making the setup and finding the company is out of 1-inch end mills. There are 0.875-inch-diameter end mills and 1.25-inch-diameter end mills, but no end mills left in 1-inch diameter. In this case, the programmer would have to change the program in order to use an end mill diameter other than the one programmed.

Most machinists will agree that the cutting tool will seldom machine the workpiece as desired on the first try. The cutting tool, workpiece, and even the machine tool itself are under a great deal of pressure during machining. The more powerful the machining operation is, the greater the pressure. Even when a milling cutter is kept quite rigid (sturdy end mill holder and short overall length), there will be some deflection of the tool during machining. This is because the cutting edge of the tool will have a tendency to push away from the surface being machined.

If you have ever scraped paint from the wall of a room, you have experienced this kind of deflection. When scraping paint, you do your best to push the scraper in a way that will remove the paint, but many times your scraper will be deflected from the surface to be scraped. In machining terms, this tendency to deflect is called Tool Pressure.

Generally speaking, the weaker the machine tool and cutting tool are, the more potential for deflection. While the small deflection may not be substantial enough to cause problems, there are times when it will. This is especially true when the accuracy required of the part is demanding.

In the previously discussed example related to milling the right side of a workpiece, even if an end mill precisely 1 inch in diameter is used, there is still the potential for deflection of the tool during machining. Depending on the expected tolerance for this surface, the amount of deflection may be enough to cause the part to be out of tolerance. If this were the case, and if fixed centerline coordinates were used, it would mean having to reprogram the milling cutter to allow for deflection.

Also note that, as the cutter dulls, deflection will increase. This means a sharp cutter will have less deflection than a dull one. This change in deflection amount during the life of a cutting tool can present real headaches while machining fixed centerline coordinates are used in the program.

3.4.3 Complex Contours

Even with the previously discussed possible problems, a case can be made for using centerline coordinates for simple parts. With the previously given simple example of milling the right side of a workpiece, it would be relatively easy to simply add or subtract the radius of the cutter to the square surface of the part to be machined. However, as the surface to be milled becomes more complicated, calculating the centerline coordinates for the path of the end mill also becomes more complicated. If angular surfaces and radii must be machined, many times it can be difficult enough to calculate coordinates on the workpiece, let alone the coordinates for the centerline of the cutter.

When the programmer calculates the tool's centerline coordinates for the tool path, he will find it very difficult indeed (Figure 3.11). In this example, it would be hard enough to calculate the actual surfaces of the workpiece and the tool's centerline coordinates.

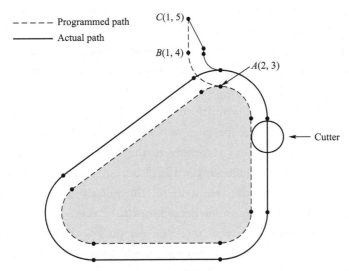

Figure 3.11 Cutter Radius Compensation Entry Moves for Material Edge Contour

The reason we present for using cutter radius compensation has something to do with roughing operations. We have already stated that, when a complex contour must be machined, it can be difficult enough to come up with workpiece coordinates, let alone cutter centerline coordinates. Moreover, there exists the complication of having to allow for a constant amount of finishing stock throughout the surface to be machined during roughing. In essence, this doubles the amount of work the programmer must do. The programmer must not only calculate the centerline coordinates of the end mill during finishing, but also calculate the centerline coordinates during the roughing operation.

3.5　CNC Acceleration/Deceleration Control

It is important to ensure smooth movement and reduce machine tools impact in high precision CNC machining. However, in practical machining, the impact and vibration on machine tools worsen cutting process and destroy machining quality. It has been reported that the unsmooth Acc/Dec control in feed movement is one of the main reasons to bring about machine tools impact and vibration.

(1) An interpreter plays the role of reading a part program, interpreting the ASCII blocks in the part program, and storing interpreted data in internal memory for the interpolator. In general, CNC issues the orders related to the interpreted data and the interpreter reads and interprets the next block while the command is being performed. However, if the time to interpret the block is longer than the time to finish the command, the machine should wait for the completion of interpretation of the next block. So a machine stop cannot be avoided. Therefore, in order to prevent machine tools from stopping, a buffer that temporarily stores the interpreted data is used. The buffer,

also called the Internal Data Buffer, always keeps a sufficient number of interpreted data and all interpreted data are stored in the buffer.

(2) An interpolator plays the role of sequentially reading the data from the internal data buffer, calculating the position and velocity per unit time of each axis, and storing the result in a FIFO (First in, First out) buffer for the Acc/Dec controller. A linear interpolator and a circular interpolator are typically used in a CNC system. A parabola interpolator and a spline interpolator are used for part of CNC system. The interpolator generates a pulse corresponding to the path data according to the type of path (e.g. line, circle, parabola, and spline) and sends the pulse to the FIFO buffer. The number of pulses is decided based on the length of path and the frequency of the pulses is based on the velocity. In a CNC system, the displacement per pulse determines the accuracy.

For example, if an axis can move 0.002 mm per pulse, the accuracy of the CNC system is 0.002 mm. In addition, the CNC system should generate 25000 pulses for the moving part to move as much as 50 mm and 8333 pulses per second to move at a speed of 1m per minute.

(3) If position control is executed by using the data generated from the interpolator, large mechanical vibration and shock occur whenever part movement starts and stops. In order to prevent mechanical vibration and shock, the filtering for Acc/Dec control is executed before interpolated data is sent to the position controller. This method is called the "Acceleration/deceleration-after-interpolation" method. An "Acceleration/deceleration-before-interpolation" method exists where Acc/Dec control is executed before interpolation.

(4) The data from an Acc/Dec controller is sent to a position controller and position control is carried out based on the transmitted data in a constant time interval. A position control typically means a PID controller and issues velocity commands to the motor driving system in order to minimize the position difference between the commanded position and the actual position found from the encoder. However, the problems of noise cannot be avoided by using an analog signal.

The acceleration, deceleration, and jerk values within the CNC or given by the CNC programmer.

(1) The linear law of acceleration and deceleration (Figure 3.12).

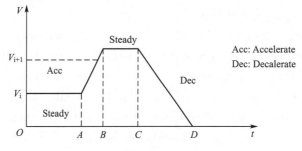

Figure 3.12　The Linear Law

(2) The exponential curve law of acceleration and deceleration (Figure 3.13).

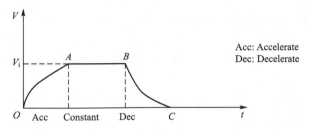

Figure 3.13 The Exponential Curve Law

3.6 PLC Function

The NCK (Numerical Control Kernel) unit, being the core of the CNC system, interprets the part program and executes interpolation, position control, and error compensation based on the interpreted part program. Finally, this controls the servo system and causes the workpiece to be machined. The PLC (Programmable Logic Control) sequentially controls tool change, spindle speed, workpiece change, and in/out signal processing and plays the role of controlling the machine's behavior with the exception of servo control.

The logic controller is used to execute sequential control in a machine and an industry. In the past, logic control was executed by using hardware that consisted of relays, counters, timers, and circuits. Therefore, it was considered as a hardware-based logic controller. However, recent PLC systems consist of a few electrical devices including microprocessors and memory (Figure 3.14). They are able to carry out logical operations, a counter function, a timer function and arithmetic operations. Therefore, a PLC system can be defined as a software-based logic controller. The advantages of PLC systems are as follows:

(1) Flexibility: The control logic can be varied by changing only a program.

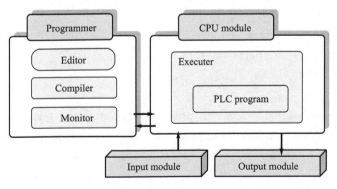

Figure 3.14 The Architecture and Function of the PLC System

(2) Scalability: The expansion of a system is possible by adding modules and changing programs.

(3) Economic efficiency: Reduction of cost is possible due to the decrease of design time. The systems are highly reliable and easily mainfained.

(4) Miniaturization: The installation dimension is smaller compared with a relay control box.

(5) Reliability: The probability of failure occurrence due to bad contact decreases because of using a semiconductor.

(6) Performance: Advanced functions such as arithmetic operations and data editing are possible.

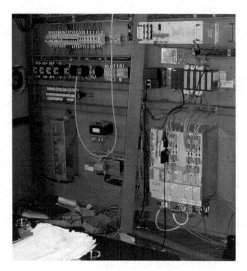

Figure 3.15 PLC

The PLC unit of a CNC system is similar to the general PLC system but there is an auxiliary controller that assists with part of the functions of the NCK unit (Figure 3.15). Therefore, the following functions are necessary:

(1) Circuit dedicated to communicating with NCK.

(2) Dual-port RAM for supporting high-speed communication.

(3) Memory for the exchanged data during high-speed communication with NCK.

(4) High-speed input module for high-speed control such as turret control.

In practice, according to the decisions of individual CNC and PLC makers, various PLC languages are used. Due to this, there is a problem with respect to maintainability and training of users. To overcome this problem, the standard PLC language (IEC1131-3) was established and usage has spread. The standard, IEC-1131-3, defines five kinds of language:

(1) Structured text (ST).

(2) Function block diagram (FBD).

(3) Sequential function charts (SFC).

(4) Ladder diagram (LD).

(5) Instruction list (IL).

Now it is necessary for users to edit programs based on the standard language and it is required for developers to implement applications for interpreting and executing a PLC program.

A programmable logic controller or programmable controller is a digital computer used for automation of electromechanical processes, such as control of machinery on factory assembly lines, amusement rides, or lighting fixtures. PLC is used in many industries and machines, such as CNC machine tools.

Basic programming of ladder logic.

As introduced by the examples of Figure 3.16, the basic ladder logic symbols are ⊣⊢ Normally open (NO) contact. Passes power (on) if coil driving the contact is on (closed). ⊣/⊢ Normally closed (NC) contact. Passes power (on) if coil driving the contact is off (open). ○ Output (or coil). If any left-to-right path of contacts passes power, the output is energized. If there is no continuous left-to-right path of contacts passing power, the output is de-energized.

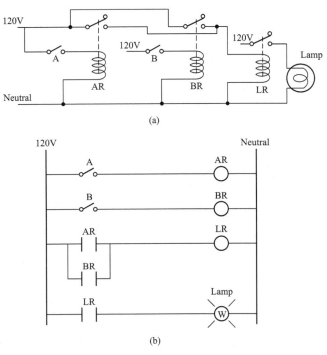

Figure 3.16　Parallel Switch Relay and Ladder Logic Circuits
(a) Equivalent relay circuit; (b) Equivalent relay ladder logic circuit

In this architecture, a CNC unit is divided into several function modules. Its hardware and software are designed with the modularization method, that is to say, each function module is made of printed circuit of same size. And the control software of function modules is also designed in modularization. Hence, customers can establish their own CNC units by combining selected function modules into the card racks of a motherboard. CNC control module, position control card, PLC card, graph display card and communication card are familiar function modules.

Exercises

1. Narrate the concept of MCU.

2. What is the function of GPU?
3. What are the main functions of control software?
4. What is interpolation?
5. What is the principle of a point by point comparison method interpolation?
6. What is cutter radius compensation? What are the advantages of the compensation?

Chapter 4 Servo System and Position Measuring Device

Objectives

- To understand the concept of servo systems.
- To understand the requirements of servo systems.
- To master the principle of position measuring devices.

4.1 Introduction

4.1.1 Concept of Servo System

Servo system is a main subsystem of the numerical control system. If we imagine that the CNC unit is the brain of a machine tool that issues "order", servo system is then the "arms and legs" of the machine tool that carries out the "order". Servo system is an important composition part of a CNC machine tool. Its performance directly affects the performances of a CNC machine tool in terms of precision, speed and reliability.

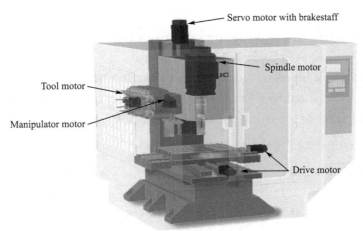

Figure 4.1 Servo System

A diagram of a typical three-axis CNC machine tool is shown in Figure 4.1. The CNC machine tool consists of mechanical, power electronic, and CNC units. The mechanical unit mainly consists of spindle assembly and feed drive mechanisms. Spindle and feed drive motors and their servo-amplifiers, power supply unit, and limit switches are parts of the power electronics group. The CNC unit consists of a computer system, position and velocity sensors for each drive mechanism. The servo system, which is introduced in this chapter, consists of mechanical, power electronic, and part of CNC unit. In the servo system, numerical commands from CNC are processed and amplified to the high voltage levels required by the drive motors. As the drives move, sensors measure their velocity and position, and the CNC unit periodically executes digital control laws that maintain the feed and tool path at programmed rates by using sensor feedback measurements.

4.1.2 Requirements for Servo System

1. Precision in CNC Positioning

For accurate machining or other processing performed by a CNC system, the positioning system must possess a high degree of precision. Three measures of precision can be defined for the CNC position system: ① control resolution, ② position accuracy, and ③ repeatability.

Accuracy is defined under the worst conditions in which the desired target point lies in the middle between two adjacent addressable points. Since the table can only be moved to one or the other of addressable points, there will be an error in the final position of the worktable.

The maximum possible positioning error is: if the target was closer to either of the addressable points, the table would be moved to the closer control point and the error would be smaller. It is appropriate to define accuracy in this worst case.

2. Repeatability

Repeatability refers to the capability of the positioning system to return a given addressable point that has been previously programmed.

This capability can be measured in terms of the location error encountered when the system attempts to position itself at the addressable point. Location error is a manifestation of the mechanical error of a positioning system, which follows a normal distribution.

3. Rapid Response and Stability

Rapid response can follow the instructions pulse quickly and play, stop, reverse frequently. It can eliminate the load disturbance as soon as possible. Rapid response is an important indicator of the dynamic quality of the servo system. It reflects the system's tracking accuracy.

Stability refers to the ability that output of a serve system oscillates as little as possible around the required output. It means the system can reach a new or restored to its original equilibrium state after a given input or external disturbances under the effect of the adjustment in short-term process. The stability of the system has a direct impact on NC machining accuracy and surface roughness.

4. A wide Speed Range

Speed range means the ratio of maximum speed and minimum speed that is required. It should be more than 10000 : 1 and hold stability.

5. Large Torque in Low Speed

The servo control of feed coordinates belongs to constant torque control. Large torque should be maintained in the entire speed range.

6. No Cumulative Error

Cumulative error is an error that gradually increases in degree or significance during a series of measurements or calculations.

4.2 Servo Systems

4.2.1 Driving Motor

The term "driving motor" is used to mean both the servo motor which moves the table, and the spindle motor which rotates the spindle. The spindle is the device that generates adequate cutting speed and torque by rotating the tool or workpiece. Consequently, high torque and high speed are very important for spindle motors and an induction motor is generally used due to the characteristics of the spindle motor. Unlike 3-phase motors, the servo motor should have characteristics such as high torque, high acceleration, and fast response at low speed and can simultaneously control velocity and position.

Machine tools, such as turning machines and machining centers, need high torque for heavy cutting in the low-speed range and high speed for rapid movement in the high-speed range. Also, motors with small inertia and high responsibility are needed for machines that frequently repeat tasks whose machining time are very short, for example, punch presses and high-speed tapping machines. The fundamental characteristics required for servo motors of machine tools are the following:

(1) To be able to get adequate output of power according to work load.

(2) To be able to respond quickly to an instruction.

(3) To have good acceleration and deceleration properties.

(4) To have a broad velocity range.

(5) To be able to control velocity safely in all velocity ranges.

(6) To be able to be continuously operated for a long time.

(7) To be able to provide frequent acceleration and deceleration.

(8) To have high resolution in order to generate adequate torque in the case of a small block.

(9) To be easy to rotate and have high rotation accuracy.

(10) To generate adequate torque for stopping.
(11) To have high reliability and long length of life.
(12) To be easy to maintain.

Servo motors are designed to possess the above-mentioned characteristics and the term comprises the DC servo motor, synchronous type AC servo motor, and induction type AC servo motor as shown in Figure 4.2.

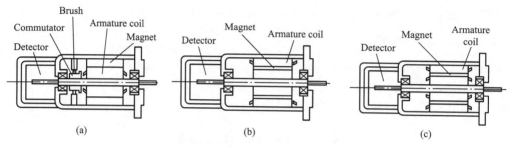

Figure 4.2 Servo Motors
(a) DC servo motor; (b) Synchronous-type AC servo motor; (c) Induction-type AC servo motor

The driving system is an important component of a CNC machine as the accuracy and repeatability depend very much on the characteristics and performance of the driving system. The requirement is that the driving system has to respond accurately according to the programmed instructions. This system usually uses electric motors while hydraulic motors are sometimes used for large machine tools. The motor is coupled either directly or through a gear box to the machine leadscrew to moves the machine slide or the spindle. Three types of electrical motors are commonly used: Stepping motor, DC servo motor and AC servo motor.

1. Stepping Motor

A stepping motor is a device that converts the electrical pulses into discrete mechanical rotational motions of the motor shaft. This is the simplest device that can be applied to CNC machines since it can convert digital data into actual mechanical displacement. It is not necessary to have any analog-to-digital converter or feedback device for the control system. They are ideally suited to open-loop systems.

However, stepping motors are not commonly used in machine tools due to the following drawbacks: slow speed, low torque, low resolution and easy to slip in case of overload. Examples of stepping motor application are the magnetic head of floppy-disc drive and hard disc drive of computer, daisy-wheel type printer, X-Y tape control, and CNC EDM Wire-cut machine.

There are four main types of stepper motors:

(1) Permanent magnet stepper (can be subdivided into "tin-can" and "hybrid", tin-can being a cheaper product, and hybrid with higher quality bearings, smaller step angle, higher power density).

(2) Hybrid synchronous stepper.

(3) Variable reluctance stepper.

(4) Lavet type stepping motor.

A sequence of a simplified stepper motor (unipolar) is shown in Figure 4.3.

Frame a——The top electromagnet 1 is turned on, attracting the nearest teeth of the gear-shaped iron rotor. With the teeth aligned to electromagnet 1, they will be slightly offset from right electromagnet 2.

Frame b——The top electromagnet 1 is turned off, and the right electromagnet 2 is energized, pulling the teeth into alignment with it. This results in a rotation of 3.6° in this example.

Figure 4.3 Stepper Motor

Frame c——The bottom electromagnet 3 is energized, and another 3.6° rotation occurs.

Frame d——The left electromagnet 4 is energized, rotating again by 3.6°. When the top electromagnet 1 is again enabled, the rotor will have rotated by one tooth position. Since there are 25 teeth, it will take 100 steps to make a full rotation in this example.

Generally step angle:

$$\theta = \frac{360}{mzk} \tag{4.1}$$

m——Phase winding.

z——Rotor teeth.

k——Single beat $k=1$, double beat $k=2$.

NC device: Sent out command pulses according to requirements; the number of command pulses represent movement distance. Change the pulse frequency can change the speed. Each pulse for motor rotates a certain angle.

Ring distribution: According to the direction of command instruction, a stepper motor steps in each phase according to the power. Ring distribution can use hardware or software.

Amplifier: Amplify ring distribution of each phase instructions, producing stepper motor drive current of each phase.

How is a stepper motor controlled?

A stepper motor performs the conversion of logic pulses by sequencing power to the stepper motor windings. Generally, one supplied pulse will yield one rotational step of the motor. This precise control is provided by a stepper driver which controls speed and positioning of the motor. The stepper motor increments a precise amount with each control pulse, converting digital information into exact incremental rotation without the need for feedback devices, such as tachometers or encoders. Since the stepper motor and driver is an open-loop system, the problems of feedback loop phase shift and resultant instability, common with servo motor systems are eliminated.

With appropriate logic pulses, stepper motors can be bi-directional, synchronous, providing rapid acceleration, run/stop, and can interface easily with other digital mechanisms. Characterized as having low-rotor moment of inertia, no drift, and a noncumulative positioning error, a stepper motor is a cost-effective solution for many motion control applications. Generally, stepper motors are operated without feedback in an open-loop fashion and sometimes match the performance of more expensive DC servo systems. As mentioned earlier, the only inaccuracy associated with a stepper motor is a noncumulative positioning error which is measured in percent of step angle. Typically, stepper motors are manufactured within a 3%~5% step accuracy.

The main difference between stepper motors and servo motors is the type of motor used and the way it is controlled. Stepper motors use between 50 to 100 pole brushless motors while the stepper motors can accurately move between step positions because of the high number of poles the motor has. Stepper motors move incrementally using pulses of current and do not require the use of a closed loop feedback system. Servo motors on the other hand require the use of a feedback system to calculate the required amount of current to move the motor.

The performance difference between a stepper and a servo is a result of the motor design. The Stepper motors have significantly higher number of poles than the servo motors have. One revolution of the stepper motor requires many current pulses through the motors windings than the servo motor. Therefore, the torque of a stepper motor is greatly reduced at higher speeds compared with the servo motor. On the other hand, the higher number of poles of a stepper motors delivers more torque at lower speeds than that of the same size servo motor. Torque reduction of a stepper motor at higher speeds can be reduced by increasing the driving voltage to the motor. Table 4.1 outlines advantages and disadvantages of both servo and stepper motors.

Table 4.1 Servo Motors Versus Stepper Motors

	Servo motors	Stepper motors
Advantages	Higher torque at high speeds	Reduced cost in drive electronics
	Reduced heat production	High torque at low speeds
		Position accuracy and repeatability
		Position stability
		High holding torque
		Easier to maintain and reliable
		Flexibility, may be used with open or closed loop configurations
Disadvantages	Drive electronics are complicated and expensive	Generates considerable heat if not using feedback
	Lower torque at low speeds	Lower torque at high speeds
		Resonance problems that must be overcome

Main control features of a stepper motor:

Step angle (θ) and Step error.

Start frequency (f_{st}).

The maximum operating frequency (f_{max}).

Acceleration and deceleration characteristic.

Start torque-frequency characteristic.

Running torque-frequency characteristic.

(1) Step angle (θ) and Step error. Step Angle is the angle of the rotor turned between two adjacent pulses. Generally, the smaller the step angle, the higher the control precision.

Step error directly affects the positioning accuracy of the implementation of the components.

When a single phase of the stepper motor gets power supply, the step error depends on sub-tooth precision of stator and the misalignment angle accuracy of each phase stator.

When multi phases of the stepper motor get power supply, it is not only the above-mentioned factors that counts but also the phase current, the performance of magnetic circuit related.

(2) The maximum starting frequency. When the motor is in idle load, it should suddenly start from stationary and enter into the steady-speed running without falling out of step. In this case, the maximum starting frequency is allowed.

When the starting frequency is higher than the ratio, the stepper motor can work. The maximum starting frequency has relation to the inertia loads of the stepper motor.

(3) The maximum operating frequency. When the frequency of the stepper motor has a continuous rise in operating time, the motor need to run without falling out of step, the max of the step frequency is called maximum operating frequency. Its value is also related to load. Obviously, maximum operating frequency is much greater than the maximum starting frequency in the same load.

(4) Torque-frequency characteristic. In the continuous operation mode, the electromagnetic torque of the stepper motor declines sharply with a higher frequency. This relationship between the torque and frequency is called torque-frequency characteristic.

How to select a stepper motor?

There are several important criteria involved in selecting a proper stepper motor:

(1) Desired mechanical motion.

(2) Speed required.

(3) Load.

(4) Stepper mode.

(5) Winding configuration.

Motion requirements, load characteristics, coupling techniques, and electrical requirements need to be understood before the system designer can select the best stepper motor/ driver/ controller combination for a specific application. Those key factors need to be considered when determining an optimal stepper motor solution. The system designer should adjust the characteristics of the elements under his/her control to meet the application requirements. Elements need to be considered include the stepper motor, driver, and power supply selections, as well as the mechanical transmission, such as gearing or load weight reduction through the use of alternative materials.

Most stepper motors are labeled as shown in Figure 4.4. The major points include the voltage, resistance and the number of degrees per step. Knowing the number of degrees per step is vital for configuring the software to properly control the machine later on. For a three-axis machine, at the very least you'll want the X and Y axis to both have identical motors. It's not the end of the world if they don't match, but it's more of a pain later on.

Figure 4.4 Label of a Stepper Motor

2. DC Servo Motor

This is the most common type of feed motors used in CNC machines. The principle of operation is based on the rotation of an armature winding in a permanently energized magnetic field. The armature winding is connected to a commutator, which is a cylinder of insulated copper segments mounted on the shaft. DC current is passed to the commutator through carbon brushes, which are connected to the machine terminals. The change of the motor speed is achieved by varying the armature voltage and the control of motor torque is achieved by controlling the motor's armature current. In order to achieve the necessary dynamic behaviour, it is operated in a closed-loop system equipped with sensors to obtain the velocity and position feedback signals.

The DC servo motor is built as shown in Figure 4.2 (a). The stator consists of a cylindrical frame which plays the role of the passage for magnetic flux and mechanical supporter, and the magnet which is attached to the inside of the frame. The rotor consists of a shaft and brush. A commutator and a rotor metal supporting frame (rotor core) are attached to the outside of the shaft and an armature is coiled in the rotor metal supporting frame. A brush that supplies current through the commutator is built with the armature coil. At the back of the shaft, a detector for detecting rotation speed is built into the rotor. In general, an optical encoder or tacho-generator is used as a detector.

In a DC servo motor, a controller can be easily designed by using a simple circuit because the torque is directly proportional to the amount of current. The factor that limits the output of the power is the heat from the inside of the motor due to current.

Therefore, efficient removal of the heat is essential to generate high torque. The velocity range of DC servo motors is very broad and the price is very low. However, friction with the brushes results in mechanical loss and noise and it is necessary to maintain the brushes.

3. AC Servo Motor (Figure 4.5)

In an AC servomotor, the rotor is a permanent magnet while the stator is equipped with 3-phase windings. The speed of the rotor is equal to the rotational frequency converter.

Figure 4.5　AC Motor

AC motors are gradually replacing Dc servo motors. The main reason is that there is no commutator or brushes in the AC servo motor so that maintenance is virtually not required. Furthermore, AC motor shave a smaller power-to-weight ratio and faster response.

The structure of AC servo motor is shown in Figure 4.6.

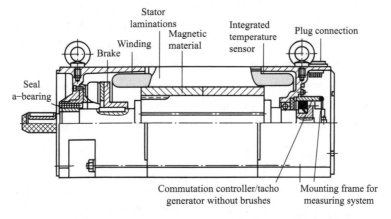

Figure 4.6　AC Servo Motor (Courtesy of Flexible Automation)

1) Synchronous-type AC servo motor

The stator consists of a cylindrical frame and a stator core. The stator core is located in the frame and an armature coil is wound around the stator core. The end of the coil is connected with a lead wire and current is provided from the lead wire. The rotor consists of a shaft and a permanent magnet and the permanent magnet is attached to the outside of the shaft.

In a synchronous-type AC servo motor, the magnet is attached to a rotor and an armature coil is wound around the stator unlike the DC servo motor. Therefore, the supply of current is possible from the outside without a stator and a synchronous-type AC servo motor is called a "Brushless Servo Motor" because of this structural characteristic. Since this structure makes it possible to cool down a stator core directly from the outside, it is possible to resist an increase in temperature. Also, because a synchronous-type AC servo motor does not have the limitation of maximum velocity due to rectification spark, a good characteristic of torque in the high-speed range can be obtained. In addition, because this type of motor has no brush, it can be operated for a long time without maintenance.

Like a DC servo motor, this type of AC servo motor uses an optical encoder or a resolver as a detector of rotation velocity. Also, a ferrite magnet or a rare earth magnet is used for the magnet which is built into the rotor and plays the role of a field system. In this type of AC servo motor, because an armature contribution is linearly proportional to torque, stop is easy and a dynamic brake works during emergency stop.

However, as a permanent magnet is used, the structure is very complex and the detection of position of the rotor is needed. The current from the armature includes high-frequency current which is the source of torque ripple and vibration.

2) Induction-type AC servo motor

The structure of an induction-type AC servo motor is identical with that of a general induction motor. If multi-phase alternating current flows through the coil of a stator, a current is induced in the coil of rotor and the induction current generates torque. In this type of AC servo motor, the stator consists of a frame, a stator core, an armature coil, and a lead wire. The rotor consists of a shaft and the rotor core that is built with a conductor.

An induction-type AC servo motor has a simple structure, which does not need the detector of relative position between the rotor and stator. However, because the field current should flow continuously during stopping, unlike the AC servo motor, a loss of heating occurs and dynamic braking is impossible.

The strengths, weaknesses and characteristics of the servo motors mentioned above are summarized in Table 4.2.

Table 4.2 Servo-motor Summary

	DC Servo Motor	Synchronous-type AC Servo Motor	Induction-type AC Servo Motor
Strengths	Low price Broad velocity range Easy control	Brushless easy stop	Simple structure No detector needed
Weaknesses	Heat Brush wear Noise Position-detection needed	Complex structure Torque ripple Vibration Position-detection needed	Dynamic braking impossible Loss of heating

(续)

	DC Servo Motor	Synchronous-type AC Servo Motor	Induction-type AC Servo Motor
Capacity	Small	Small or medium	Medium or large
Sensor	Unnecessary	Encoder, resolve	Unnecessary
Life length	Depends on brush life	Depends on bearing life	Depends on bearing life
High speed	Inadequate	Applicable	Optimized
Resistance	Poor	Good	Good
Permanent magnet	Exists	Exists	None

4. Linear Motor (Figure 4.7)

A linear electric motor is an AC rotary motor laid out flat. The principle of producing torque in rotary motors is applicable to producing force in linear motors.

Figure 4.7　Linear Motor

Despite of the electromagnetic interaction between a coil assembly and a permanent magnet assembly, the electrical energy is converted to linear mechanical energy to generate a linear motion. As the motion of the motor is linear instead of rotational, it is called linear motor. Linear motors have advantages of high speeds, high precision and fast response. In the 1980s, machine tool builders started using linear motors with the common motion control servo drives in the machine tool design.

Among different designs of linear motors, permanent magnet brushless motors demonstrate a high force density, high maximum speed, and stable force constant. The lack of a brushed commutator assembly has advantages of fewer maintenance, higher reliability and better smoothness.

An ironcore brushless linear motor (Figure 4.8) is similar to a conventional brushless rotary motor slit axially which is then rolled out flat. The unrolled rotor is a stationary plate consisting of magnets tiled on an iron back plate and the unrolled stator is a moving coil assembly consisting of coils wound around a laminated steel core. Coil windings are typically connected in conventional 3-phase arrangement and commutation is often performed by Hall-effect sensors or sinusoidal. It has high efficiency and good for continuous force.

An ironless linear motor consists of a stationary U-shaped channel filled with permanent magnets tiled along both interior walls. A moving coil assembly traverses between two opposing rows of magnets. Commutation is done electronically either by Hall-effect sensors or sinusoidal. The ironless linear motor has the advantages of lower core mass, lower inductance and no cogging for smooth motion as the ironless motors have no attractive force between the frameless components.

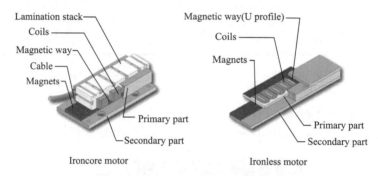

Figure 4.8 Ironcore and Ironless Linear Motor

4.2.2 Position Control

1. Speed Control

The speed of the motor is measured and compared with the reference signal. Usually the speed is determined as a time derivative of the rotor angle. The signals are forwarded to the speed control system, which produce reference value. With other applicable signals from various parts of the drive, these are employed in the flux and torque control system.

2. Velocity Feedback Device

The actual speed of the motor can be measured in terms of voltage generated from a tachometer mounted at the end of the motor shaft. DC tachometer is essentially a small generator that produces an output voltage proportional to the speed. The voltage generated is compared with the command voltage corresponding to the desired speed. The difference of the voltages can be used to actuate the motor to eliminate the error.

Tacho-generators (Figure 4.9) are AC or DC generators that output a voltage in proportion to the rota-

Figure 4.9 Tacho-generator

tional speed of a shaft on a rotating electrical machine (electric motor), and thus are used to measure the speed and direction of rotation.

3. Position Control

Position control requires an unusually high level of operating performance. The motor starts, runs and stops through turning a specific total angle. Position control drivers are widely used in manufacturing industry, such as elevators, material handling equipment, packaging systems, and processing lines besides machine tools. An angular position sensor, usually an encoder, provides the control feedback. The position control is the main outer loop, with speed and torque loops as inner ones. The position controller can be linear (PI or PID), fuzzy or neural-fuzzy, or of variable structure type.

4.3 Requirements and Classifications for Position Measuring Devices

In a closed-loop system, information about the output is fed back in comparison with the input. Thus, position-measuring devices are required to close the loop. Position measuring devices are usually rotary or linear devices, which provide position and velocity signals.

1. Feedback Device

In order to have a CNC machine operating accurately, the positional values and speed of the axes need to be constantly updated. Two types of feed back devices are normally used, positional feed back device and velocity feed back device.

2. Positional Feed Back Device

There are two types of positional feed back devices: linear transducer for direct positional measurement and rotary encoder for angular or indirect linear measurement.

A linear transducer is a device mounted on the machine table to measure the actual displacement of the slide. In such a way that backlash of screws and motors, etc., would not cause any error in the feed back data. This device is considered to be of higher accuracy and more expensive in comparison with other measuring devices mounted on screws or motors.

4.3.1 Requirements for Position Measuring Devices

To determine the slideway position of each axis of a machine tool, position-measuring device is required, which monitors and compares the present position with the command position for every movement of the axis.

These are requirements for selecting position-measuring devices in CNC machine tools.

(1) There should be maximum reliability and minimum interference with the quantity being measured.

(2) The requirements are satisfied for high accuracy and speed range.

(3) The installation and maintenance of the measuring devices should be easy. Devices are adaptable to the working environment.

(4) The cost to purchase should be low.

4.3.2 Classifications for Position Measuring Devices

Measuring devices can be classified according to following considerations:

(1) Direct and indirect measurement.

(2) Incremental and absolute measurement.

(3) Digital and analogue measurement.

1. Direct and Indirect Measurement

Direct measuring system is a system with the position measuring devices directly coupled to the machine slideway or table to be positioned. It is independent of the leadscrew or drive element (and hence obviates errors associated with the leadscrew or drive element). The disadvantage of high relative cost exists because accuracy must be maintained throughout measuring length. It has a further disadvantage of placing the full mechanical resonance of the machine inside the feedback loop. Direct measurement can be effected with linear measuring devices such as optical grating or inductosyn. Invariably, though using linear devices is a costly option as the demand for higher accuracy components increases they are becoming more popular.

Indirect measuring system is a system with the position measuring devices mounted on the leadscrew or drive element of the machine member to be positioned. Indirect measurement has advantages of lower cost, simpler mounting on the machine tool and easier maintenance. Accuracy requirements can be reduced by backlash, pitch error in the leadscrew and fine devices. Indirect systems are usually fitted with rotary position measuring devices, e.g. resolver and encoder.

2. Incremental and Absolute Measurement

In an incremental system, each displacement is determined by its departure from being measured as the increase in distance from the preceding position. Incremental system has the disadvantages that the position can only be determined relatively and a previous measurement error influences all following measurements (cumulative error).

In an absolute system all coordinates are measured from a fixed datum position or center of coordinates without reference to previous coordinates. Restart after power supply breakdown (midway start) is easier, and accumulated error is less. The digital absolute measuring device is expensive for a high resolution.

3. Digital and Analogue Measurement

In a digital measuring system, position or displacement is measured by the use of discrete values. Digital measuring systems can be incremental (e.g. with optical grating and incremental encoder) or absolute (e. g. absolute encoder). Digital signals are easier to store, more reliable to transmit and error-free to reproduce. Digital measuring systems require no separate analog/digital converter.

In an analogue measuring system, the displacement is converted into another physical analog which can be easily measured. The possible physical analogs may be extremely varied in nature. An analog position-measuring device may be an inductive transducer whose output is continuously proportional to displacement (e. g. resolver and inductosyn).

The analogue system implies that a signal such as electrical voltage magnitude will represent a physical axis position, or to say it another way. We use the physical variable of distance to represent a voltage. A specific slideway displacement will be "analogous" with an induced voltage. Analog devices are generally simpler, more robust and less expensive. However, the measuring process being continuous, it is more difficult to provide display of the moving member. Analog displacement-measuring devices are suitable for small-and medium-sized CNC machine tools.

4.4 Position Measuring Devices

4.4.1 Resolver

A resolver (Figure 4.10) is a position sensor or transducer which measures the instantaneous angular position of the rotating shaft to which it is attached. Resolvers and their close cousins, synchros, have been in use since World War II in military applications, such as measuring and controlling the angle of gun turrets on tanks and warships. Resolvers are typically built like small motors with a rotor (attached to the shaft whose position is to be measured), and a stator (stationary part) which produces the output signals.

All resolvers produce signals proportionally to the sine and cosine of their rotor angle. Since every angle has a unique combination of sine and cosine values, a resolver provides absolute position information within one revolution (360°) of its rotor. This absolute (as opposed to incremental) position capability is one of the resolver's main advantages over incremental encoders.

Figure 4.10 Resolver

4.4.2 Optical Grating

The optical grating (Figure 4.11) consists of two transparent scales made from glass. They require delicate handling (photo-etched, or vapor deposition) a series of parallel lines closely and uniformly spaced together. The transparent and black lines can be equal in width. One of them is long while the other is short. Generally, the long scale is called scale grating and mounted on the moving member. The short is called index grating and mounted on the stationary member.

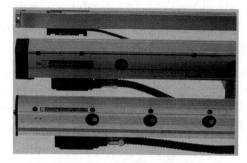

Figure 4.11 Optical Grating

As shown in Figure 4.12, an optical rating measuring system consists of a light source one collimating lens, optical grating and photocells. The long fixed scale grating extends over the length of the machine tool's slideway travel, with a short index grating overlaying it. Except scale grating, other units are held in a reading head.

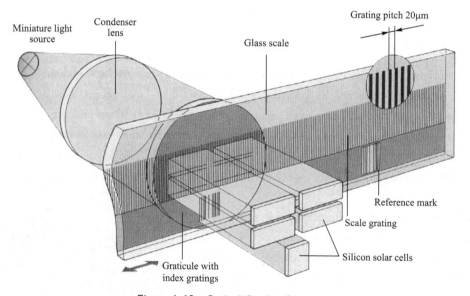

Figure 4.12 Optical Grating Structure

The optical grating principle is based upon the well known "Moire fringe effect". The lines of both scales are tilted against each other by a very small angle θ. This causes an interference effect between them at the intersection of the lines. This pattern is known as a Moire fringe.

In practice, when the slideway is moved, the lines on the two scales are displaced. As a result, the fringe pattern travels at right angles across them with their direction of movement being dependent on the right (plus) or left (minus) direction of the slideway. As shown in Figure 4.13, a dark (light) fringe pattern is produced across the width of the grating and moves in the sequence shown in the illustrations.

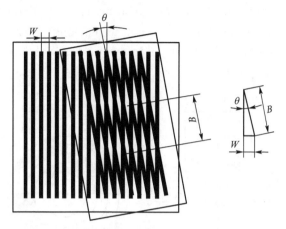

Figure 4.13 Moire Fringe Effect

W is the width of Moire fringe, B is grid distance.

$$W = \frac{B}{\sin\theta} \tag{4.2}$$

θ is very small, $\sin\theta \approx \theta$.

$$W \approx \frac{B}{\theta} \tag{4.3}$$

Suppose $B=0.01\text{mm}$, $\theta=0.01$ radian, Then $W=1\text{mm}$. It means magnification factor is 100.

4.4.3 Encoder

Encoders (Figure 4.14) can measure angular position or displacement. They can be classified into two primary types:

① Absolute encoders that produce a code value, which represents the absolute position directly.

② Incremental encoders that produce digital signals which increase or decrease the measured value in incremental steps.

The rotary encoder is a device mounted at the end of the motor shaft or screw to measure the angular displacement. This device cannot measure linear displacement directly, so that error may occur due to the backlash of screw and motor etc. Generally,

Figure 4.14 Encoders

this error can be compensated by the machine builder in the machine calibration process.

The absolute encoder (Figure 4.15) is a kind of revolving measurement device. In the absolute encoder, the value of the actual position is immediately measured when the system is switched on. Absolute encoders do not need a counter since the measured value is derived directly from the graduation pattern. In most cases, the output from the encoder is in the form of a pure binary code or in gray code.

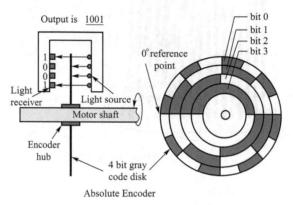

Figure 4.15 Absolute Rotary Encoder

Absolute encoders have three types: brush, optical and magnetic.

An incremental encoder is a pulse generators counter. Incremental encoders are the simplest position-measuring devices but do not give an unambiguous indication of the position as absolute shaft encoders do. The information is given only relatively. The impulses have to be counted. As is shown in Figure 4.16, the encoder contains a transparent disk mounted on the shaft and marked with a precise circular pattern of alternating clear and opaque segments on its periphery. A fixed source of light is provided on one side of the disk, and one photocell is placed on its other side.

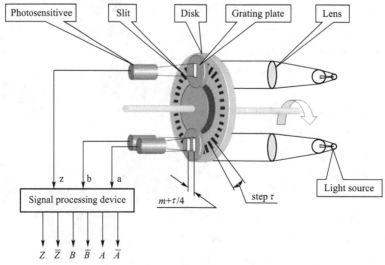

Figure 4.16 Incremental Encoders

When the disk rotates, light is periodically permitted to fall on the photocell, which consequently produces a sinusoidal output signal with slow rising and falling edges in the millivolt range. This signal is amplified and fed to a shaping circuit, which converts it to a square wave. The square wave is changed into short-duration pulse by differentiating element. A pulse is generated each time a line breaks the light circuit. The number of pulses generated in per revolution is a function of the number of lines etched on the disk and called resolution. Typical disk may contain from 200 to 18000 lines.

Incremental encoders are shaft-driven devices delivering electrical pulses at their output terminals. The number of pulse decides distance. The pulse count can be used to determine the linear axis position of the workable by factoring in the leadscrew pitch and the gear.

Exercises

1. State the composition of servo systems.
2. What are requirements for servo systems?
3. What is the principle of open-loop control of the stepper motor?
4. How to select stepper motors?
5. What is the principle of the AC servo motor?
6. Narrate the principle of optical grating.
7. Narrate the principle of incremental encoders.

Chapter 5   Mechanical Construction and Tool System of CNC Machines

Objectives

- To understand the structure of CNC machine tools.
- To understand the CNC tool system.
- To master selecting method of cutter tools.

5.1   CNC Machine Tools

The purpose of a machine tool is to cut away surplus material. Usually the materials are supplied to leave a workpiece of the required shape and size, producing an acceptable degree of accuracy and surface finish. The machine tools should possess certain capabilities in order to fulfill these requirements. It must be:

(1) Able to hold the workpiece and cutting tool securely.

(2) Endowed the sufficient power to enable the tool to cut the work piece material at economical rates.

(3) Capable of displacing the tool and workpiece relative to one another to produce the required workpiece shape. The displacements must be controlled with a degree of precision which will ensure the desired accuracy of surface finish and size.

Early machine tools were designed so that the operator could stand in front of the machine while operating the controls. This design is no longer necessary since the operator no longer controls CNC machine tool movements. On conventional machine tools, only about 20% of the time was spent removing materials. With the addition of electronic controls, actual time spent removing metal has increased to 80%, even higher. It has also reduced the amount of time required to bring the cutting tool into each machining position.

5.1.1   CNC Lathe

The engine lathe, one of the most productive machine tools, has always been an efficient means of producing round parts.

Features of CNC Lathe:

(1) The tool or material moves.

(2) Tools can operate in 1 − 5 axes.

(3) Larger machines have a MCU which manages operations.

(4) Movement is controlled by motors.

(5) Feedback is provided by sensors.

(6) Tool magazines are used to change tools automatically.

Tools:

(1) Most are made from high speed steel (HSS), tungsten carbide or ceramics.

(2) Tools are designed to direct waste away from the material.

(3) Some tools need coolant, such as oil, to protect the tools.

Tool Paths, cutting and plotting motions:

(1) Tool paths describes the route the cutting tool takes.

(2) Motion can be described as point to point, straight cutting or contouring.

(3) Speeds are the rate at which the tool operates, e. g. rpm.

(4) Feeds are the rate at which the cutting tool and workpiece move in relation to each other.

(5) Feeds and speeds are determined by cutting depth, material and quality of finish needed. e. g. harder materials need slower feeds and speeds.

(6) Rouging cuts remove larger amounts of material than finishing cuts.

(7) Rapid traversing allows the tool or workpiece to move rapidly when no machining is taking place.

1. Purpose of CNC Lathe

Both the CNC lathe and the common lathe are mainly used for machining the revolving body parts such as file axes and plates. However, compared with the common lathe, the machining accuracy of the CNC lathe is higher. Its machining quality is more steady and efficiency is higher. The suitability is stronger and the working strength is lower. Especially, the CNC lathe is fit for machining some complex-shaped parts like axes and plates.

2. Conventional Cutter and Fixture of CNC Lathe

The turning tools of the CNC lathe are similar to those of the common lathe, mainly including the welding mode and the mechanical clamping mode. In the CNC turning machining, the usual shaped tools contain small radius circular turning tools, non-rectangular slotting tools and thread turning tools, etc. But these shaped tools should be used rarely in the practical application that is occasionally used or not used at all.

However, if it is necessary to use them, the detailed instruction should be given in the technological file and the machining program list.

As for the fixtures of CNC lathe, not only the general three-jaw centering chuck and four-jaw centering chuck are usually used, but the auto-control hydraulic, electric and air fixtures are used very often in mass producing. In addition, the other appropriate fixtures are usually applied to the machining of the CNC lathe, and are used for axes parts and plate parts.

1) Axis parts fixtures

Axis parts fixtures have an automatic clamp chuck, a center, a three jaw chuck and a rapidly adjustable universal chuck, etc. When the CNC lathe is machining the axes parts, the stocks clamp between the spindle center and the tailstock center, and the poking plate center on the spindle promotes swiveling around. These fixtures can transmit a big enough torque when round turning as to fit the rapid swiveling turning of the spindle.

2) Plates parts fixtures

The fixtures used for the plate parts machining have adjustable dog mode chuck and rapid adjustable chuck. These fixtures are fit for the mode CNC lathe without tailstock chuck.

3. CNC Lathe Tool Compensation

CNC lathe often needs some different kinds of tools to machine a part. But when each tool is machining a part, the tool nose location is not the same. In fact, the tool compensation is just to measure the location difference of each tool, and unify the tool rose of each cutter on some clamped location of the same work coordinate system. Then, each tool nose can move on the coordinate which is the same work coordinate system of addresses. Here are some common tools compensation methods for CNC lathe:

(1) Automatic tool compensation.

(2) Trial cutting. The trial cutting is mainly used for the CNC lathe which closed-loop or open-loop controls.

(3) Practice tool compensation inside the machine. Tool compensation inside the machine is to touch a fixed touch head with cutters, measure the tool deflection and correct it.

(4) Practice tool compensation by reference point location.

Set a function or adjust the mechanical stopper location of each coordinate axis of the machine bed with the CNC system parameter, and set up the reference points on the tool compensation reference points which correspond with the tool starting points. In this way, when the machine bed returns back to the reference point of the staring operator, it can make the tool nose return to its starting position.

4. Components of CNC Lathe

The CNC lathe is a machine tool that can get rid of the materials from workpiece blank which is clamped on the axis and rotated about it. This machine tool is primarily used for machining surfaces and end faces of revolution parts of all metal cuttings. Most of them are down with a sharp single-point cutting tool. CNC lathes use turret to hold cutting tools rigidly and move them accurately.

In addition, the turret also has all automatic tool changing function. It is used to quickly take out an old tool and put a new tool into its cutting position. A front turret built to move tools from below the spindle centerline up to the workpiece. A rear turret, on the other hand moves tools from above the spindle centerline down to the workpiece, so machines equipped with front and rear turrets can execute the cutting operation from above and below the workpiece simultaneously.

The following is the components of the CNC lathe (Figure 5.1).

Mechanical Construction and Tool System of CNC Machines Chapter 5

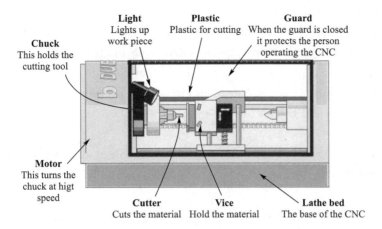

Figure 5.1 CNC Lathe

1) Vice

It holds the material to be cut or shaped. Material must be held securely. Otherwise, it may "fly" out of the vice when the CNC begins to machine. Normally, the vice will be like a clamp that holds the material in the correct position.

2) Guard

The guard protects the person using the CNC. When the CNC is machining the material small pieces can be "shoot" off the material at high speed. This could be dangerous if a piece hit the person operating the machine. The guard completely encloses the dangerous areas of the CNC.

3) Chuck

The chuck's function effect is to connect the spindle and to clamp the workpiece. It holds the material that is to be shaped. The material must be placed in it very carefully so that when the CNC is working the material is not thrown out at high speed.

4) Motor

The motor is enclosed inside the machine. This is the part that rotates the chuck at high speed.

5) Lathe bed

Lathe bed is the base of the machine. Usually a CNC is bolted down so that it cannot move through the vibration of the machine when it is working.

The lathe bed is the main frame, which supports all the components. It provides a path for chips when they fall away. The purpose of the slanted lathe bed is to allow chips to fall down easily. It also has guide ways to lathe carriage to slide lengthwise easily. The height of the lathe should be appropriate to the technician to do his job easily and comfortably.

6) Cutting tool

Cutting tool is usually made from high quality steel and it is the part that actually cuts the material to be shaped.

7) Headstock

The headstock contains a spindle and a changeable speed gear, which provides a number of different spindle speeds. The spindle is driven through the gearbox.

8) Carriage

Its function is mainly to move the cutting tool into the revolving work. It contains the saddle, the cross-slide and the apron.

9) Turret

Its main function is to hold the cutting tool and replace an old tool (with a new tool during a tool change).

10) Tailstock

The function of the tailstock is mainly to support the right function end of the workpiece.

11) MCU

MCU consists of electronics and a control hardware that reads and interprets the program of instruction. Then MCU convert it into mechanical motion of the machine tool or other processing equipment. In fact, MCU is a computer used to store and process the CNC programs input.

Several different cutting tools are used to produce a part when required. The used tools must be replaced quickly for the next machining operation. Therefore, most of the CNC machine tools are equipped with automatic tool changers. For example, the machining centers are equipped with tool magazines and the CNC lathe is equipped with turrets. On most machines, with automatic tool changers, the turret can rotate in any directions, forwards or backwards. On receiving a tool change command from the MCU, the turret moves to a safe tool change position, takes the old tool out and puts a new one into it. Then the turret proceeds to move to the proper coordinates programmed for cutting the part with new tools. As far as the turning center equipped with the programmable tailstocks is concerned, the tailstock must be first moved back before a tool change is executed.

5.1.2 CNC Machining Centers

Milling is the most versatile machining process. Metal removal is accomplished through the relative motions of a rotating, multi-edge cutter, and multi-axis movement of the workpiece. Milling is a form of interrupted cutting where repeated cycles of entry and exit motions of the cutting tool accomplish the actual metal removal and discontinuous chip generation. The milling machine has always been one of the most versatile machine tools used in industry. Operations, such as milling, contouring, gear cutting, drilling, boring, and reaming, are only a few of the many operations which can be performed on a milling machine. Machining center is one of milling machines.

All milling machines, from compact tabletop models to the standard vertical knee mill and the massive CNC machining centers, operate on the same principles and operating parameters. The most important features of these operating parameters are:

(1) Cutting speed: the speed at which the tool engages the work.

(2) Feed rate: the distance the tool edge travels in one cutter revolution.

(3) The axial depth of cut: the distance that the tool is set below an unmachined surface.

(4) The radial depth of cut: the amount of work surface engaged by the tool.

The capabilities of the milling machine are measured by motor horsepower which determines maximum spindle speeds and spindle taper size.

A machining center is a machine for both milling and whole making on a variety of non-round or prismatic shapes. The unique feature of the machining center is the tool changer. The tool changer system moves tools from storage to spindle and back again in rapid sequence. While most machining centers will store and handle 20 to 40 individual tools, some will have inventories of over 200.

The heart of the milling operation is the milling cutter. These are rotary tools with one or more cutting edges, each of which remove just a small amount of material as it enters and exits the workpiece. The variety of cutter types is almost limitless. One of the more basic is the face mill cutter used for milling flat surfaces. Used at high speeds, they range from three inches to up to two feet in diameter. Some face mills will simultaneously mill a shoulder that is square to the surface.

Work that requires edge preparation, shoulders, and grooves, is accomplished with other cutting mills. An end mill cutter is a tool with cutting edges on its end as well as on its periphery. End mills are used for short, shallow slots and some edge finishing. Circular grooving or slotting cutters are more adapted to the making of longer and deeper slots. This is because end mills are more easily deflected during heavier cuts. Chamfers and contour milling are performed with specially shaped end mills.

In all kinds of milling, a critical component is the workholding device, which has the ability to be changed over quickly to present new work or work surfaces to the tooling. Machining centers can utilize long machine beds, pallet changers and multi-storied "tombstone" part holders to enable new work to be set up and positioned while previously setup workpieces are being milled.

Machining centers (Figure 5.2) can also incorporate two useful accessories. One is the touch-trigger probe which, with its computer software, will dimensionally check workpiece measurements before removal from the machining center. The probe is stored with other tooling for quick application. The second accessory is the tool presetting machine which allows the technician to assemble the tooling according to the programmed part requirements before placing tools in the machining center tool storage.

Figure 5.2 CNC Machining Center

Machining centers are CNC machine tools equipped with automatically tool change device and tool magazines. They have been improved by increasing tool magazines and rotation worktable on the basis of a CNC milling machine. Therefore, machining centers have functions of milling, boring, drilling and so on. The main characteristics of a CNC machine center is as follows:

(1) There is a high centralized working procedure. Once a part is held, the machining for many surfaces can be finished.

(2) It can be equipped with automatically dividing unit (or rotation worktable) and tool magazine system.

(3) It can automatically change the spindle speed. Feed quantity and motion path of the cutter are relative to the workpiece.

(4) Its productivity is five or six times higher than the common CNC machine. It is especially applied to machining parts which are complex-shaped, having higher precision and frequent changing in variety.

(5) The operator's labor intensity is very low, while the machine's structure is complex, and the operator's technology level should be high.

(6) Machine's cost is high.

For different kinds of machining centers, their components are mainly composed of general parts, spindle parts, CNC systems, automatically tool changing systems, and some accessories.

Machining centers may either be vertical or horizontal. There is also a universal type capable of both orientations. The vertical type is often preferred when work is done on a single face. With the use of rotary tables, more than one side of a workpiece, or several workpieces, can be machined without operator intervention. Vertical machining centers using a rotary table have four axes of motion. Three are lineal motions of the table while the fourth is the table's rotary axis.

The classification of machining centers is as follows:

(1) Vertical machining center. Its spindle axial line is vertical. It is applied to machining plate parts. The rotation workable can be mounted on the level worktable to machine helical line.

(2) Horizontal machining center. Its spindle is horizontal, and is equipped with dividing rotation table, inclining three-five motion coordinates. It is applied to machining the box body parts.

(3) Planer machining center. Its spindle is usually vertical, and is equipped with changeable spindle head accessories. It is applied to big-sized or complex-shaped parts.

(4) Multipurpose machining center, with both horizontal and vertical function. Once a workpiece is held, the machining for all the surfaces can be performed except for the external face. The multipurpose machining center can reduce the configuration error. It can also avoid the second holding. It has high productivity and low cost.

(5) Machining center with mechanics and tool magazine. The tool changing device of

the machining center is commonly performed by the tool magazine and mechanics. The advantage of this type of machining center has a very wide machining range.

(6) Turret magazine machining center. This type of machining center is widely used on the basis of the small-sized vertical center.

In order to obtain high accuracy and repeatability, the designing and making of the machine slide and the driving leadscrew of a CNC machine is of vital importance. The slides are usually machined to reach high accuracy and coated with anti-friction material, such as PTFE (polytetrafluoroethylene) and Turcite, in order to reduce the stick and slip phenomena. Large diameter recirculating ball screws are employed to eliminate the backlash and lost motion.

Other design features such as rigid and heavy machine structure; short machine table overhang, quick change tooling system, etc. also contribute to the high accuracy and high repeatability of CNC machines.

5.2 Main Structure of CNC Machine Tools

The mechanical system is composed of the frame section, the drive section, and the guide system. This section will provide detailed description of each of these.

1. Drive

The drive mechanics of CNC machines convert torque provided by the electric motors into linear motion of the tool head. Screws with threaded nuts provide a simple and compact way to transmit this power. A ball screw and a ball nut system will be used because of its low friction and high efficiency. ACME (Association of Consulting Management Engineers) screws will not be used because neither of their advantages, a larger weight supporting capacity nor the simplicity of self locking, have application for this machine. Instead, the ability to reduce the required torque needed to produce the specified linear speeds, due to the fact that ball screws operate with a superior efficiency, makes the ball screws the obvious choice for all three axes. In addition, the lack of heat generation caused by friction and an increased reliability support the decision to implement ball screws and ball nuts as a means of power transmission.

2. Guide

The guide rails support the weight of the gantry and tool head while providing the alignment during the movement of the gantry. The linear rod guide rails will be case hardened steel shafts with ball bushings. However, a more complex shaft or support rail may be required if the weight and loads on the gantry create deflections above the specified tolerances of the machine.

3. Frame

The machine frame is divided into the gantry sides and the base table. CNC frame materials need to have some strength in order to support the weight of the gantry, and the cutting head as well as withstand forces resulting from the milling process.

Stiffness is also required to prevent any deflections due to both static forces and dynamic forces resulting from the acceleration of the tool head. Weight is important because the mass of the frame contributes to both the static and acceleration forces. The best frame material should accomplish all three requirements(strength, stiffness and weight), offer excellent machinability, and be available at a low cost. Reviewing of the materials, we can see that high density polyethylene (HDPE) offered the best combination of the above five selection factors. Among the plastics it is one of the least expensive choices. Also, it will be very easy to machine, while still providing sufficient strength and rigidity.

4. Gantry Sides

The gantry sides support the weight of the upper gantry and the head while traveling on the lower guide rails. HDPE will be used to create the gantry sides, because the gantry must be light-weight to reduce inertia forces during acceleration.

5. Base Table

The base table will support the material to be worked on, and act as the base of the machine. Constructing the base table will require a large amount of material and involve a great deal of machining and assembly. Because HDPE is low cost and it can be easily machined, it will be used for the base. In this application, the strength and stiffness qualities of HDPE will be tested. The weight supported by the lower guide rails, the rails that allow the machine to move along the length of the table, might create excessive deflections in the sides of the base table. This can cause a displacement of the tool head that exceeds the design tolerances. It is believed that HDPE will be able to withstand these forces and maintain tolerances. If deflections become an issue, a hybrid system will be considered, in which aluminum will be used to reinforce the HDPE base sides.

6. Chip Conveyor (Figure 5.3)

Chip conveyor is a moveable belt that helps to remove metal chips produced by ma-

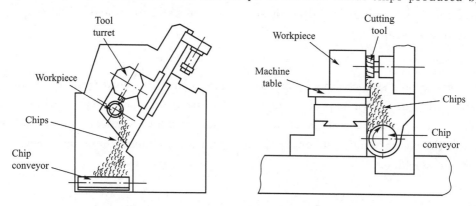

Figure 5.3　Chip Conveyors Used in CNC Machine Tools

chine tools. On some lathe, the chip conveyor is placed inside the coolant tank. Chip conveyor usually use heavy-duty steel link belts arranged in a serpentine shape.

5.2.1 Spindle Design

Everyone knows the importance of the basic spindle specs when he/she purchases a CNC machining center: max spindle speed, peak spindle motor horsepower and max spindle motor torque.

The spindle is the workhorse of your machining center. Make sure the manufacturer of your machining center has taken spindle design seriously and has invested in quality components that will help increase the longevity of your machining center's spindle (Figure 5.4).

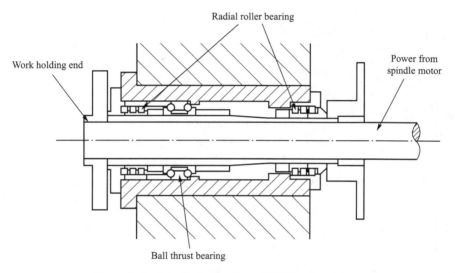

Figure 5.4 Spindle Design for a CNC Turning Center

The power you need depends upon the stock you're cutting. Obviously torque, speed and horsepower are important specs to evaluate when you're getting ready to purchase a vertical machining center, but there are a few other considerations that will be critical to the overall performance of your spindle and your overall satisfaction with the investment you make when purchasing a CNC vertical machining center. Since many integral parts of the spindle are buried within it, finding out what you need to know requires preparation, research and asking the right questions.

1. What's Inside and Why It Matters

At first blush, it might appear that the actual spindle doesn't do much on a vertical machining center, the tool cuts the metal, the table moves, the motion control system controls the precision and movement, and the software does the rest. The spindle is nothing more than a motor that lets a tool be attached to it and takes commands from a servo.

The above depiction may be true. The spindle may not be overly complex, smart or beautiful, but it sure does work hard and endures a lot of abuse. The amount of force for wear and tear it must endure means the design of the spindle and the quality of the parts buried within the spindle is vital to your spindle's performance and its lifespan. The spindle is truly the heart of the machining center.

Quality components not only determine the longevity of the spindle, but also determine how the spindle handles speed, torque and vibration. When you start to research spindle technology, you'll find the bearing system is often the focus of the discussion. An overview of key considerations when researching a CNC machining center's bearing system covers the highlights—material, type, arrangement and lubrication.

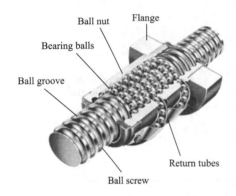

Figure 5.5　Ball Screw Structure

2. Bearing System of the Spindle (Figure 5.5)

In a bearing system, the balls roll between the inner and outer steel raceways. The material used for the ball bearings affects temperature, vibration levels and the life of the spindle. Hybrid ceramic bearings offer distinct advantages over typical steel ball bearings.

1) Advantages of hybrid ceramic bearings

Ceramic ball bearings have 60% less mass than steel balls. This is significant when a ball bearing is operating, particularly at high rotational speeds, centrifugal forces push the balls to the outer race and even begin to deform the shape of the ball. When the bearing starts to deform, it starts wearing faster and leads to deterioration. Ceramic balls with less mass will not be affected as much at the same speed. In fact, the use of ceramic balls allows up to 30% higher speed for a given ball bearing size without sacrificing bearing life, according to information from a manufacturer of high-speed milling spindles.

2) Elimination of cold welding

There is no metal to metal contact and ceramic balls do not react with steel raceways, again eliminating micro or cold welding and associated adhesive and abrasive wear. Cold welding occurs when the microscopic cold welding of ball material to the raceway causes surface wear. The cold welds actually break as the bearings rotate and that creates surface roughness that leads to heat generation and bearing failure.

3) Operate at lower temperatures

Due to the nearly perfect roundness of the ceramic balls, hybrid ceramic bearings operate at much lower temperatures than steel ball bearings, which results in longer life for the bearing lubricant.

4) Vibration levels are lower

Tests show that spindles utilizing hybrid ceramic bearings exhibit higher rigidity and have higher natural frequencies, making them less sensitive to vibration, which results in longer life for the bearing lubricant.

5) Types of bearings

There are also different types of bearings, with angular contact ball bearings being the type most commonly used in high-speed spindle design (Figure 5.6). These bearings provide precision, load carrying capacity, and the speed needed for cutting metal. The precision balls are fitted into a precision steel race and provide both axial and radial load carrying capacity.

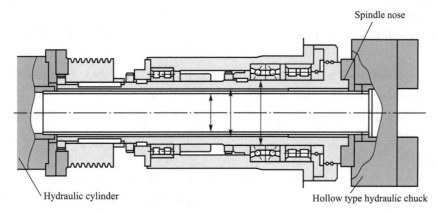

Figure 5.6 Spindle of Numerically Controlled Lathe

The other type of bearings sometimes used in spindles includes taper roller or cylindrical roller bearings. The roller (or cylindrical) bearings offer higher load carrying capacity and greater stiffness than ball bearings, and are used in spindles with specific RPM (Resolutions per Minute) requirements and applications. Usually, the spindle manufacturer will use both types in different parts of the spindle-dependent upon the type of load the bearing must counteract.

6) Lubrication

Proper lubrication of bearings is essential. There are several systems that machine tool manufacturers use to keep the bearings properly lubricated, such as oil-mist, oil-air, oil-jet and pulsed oil-air.

Such systems are sometimes necessary if bearing spindle speeds are in excess of 18,000 rpm, but they add maintenance cost and increase replacement cost of the spindle. Additionally, these lubrication systems must be monitored to make sure the ratio and the amount of oil and air and/or mist are correct.

Permanently lubricated bearings are the best option for keeping maintenance costs down and replacement costs lower. With permanently lubricated bearings, you don't have to hassle with lubrication. It is handled during the assembly of the spindle. The bearings can also be pre-packed with grease (permanent lubrication) by the bearing suppliers.

3. Types of Spindles

Spindle technology offers various ways to drive the spindle-belt, gear, inline, and

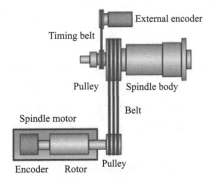

Figure 5.7　Spindle Driving Mechanism

built-in motor. With a belt-driven spindle, make sure the belt is easy to maintain and easily accessible to minimize maintenance costs. Additionally, the type of belt will affect the noise level of the machine. A belt with a herringbone design is quieter than other belt designs because of the way that it disperses trapped air to reduce the noise (Figure 5.7).

Gear-driven spindles add to the cost of the machine and can be noisier and require more maintenance than their competitor, the belt-driven spindle. There was a time when a gear-driven spindle may have been preferred over a belt-driven spindle, but the advances in materials and belt design have proven it to be a low maintenance alternative.

The inline spindle (sometimes called Direct-drive Spindle) is designed so the spindle is coupled directly to the motor. Inline spindles provide excellent surface finish, smoother and quieter operation.

Another type of spindle is the built-in motorized spindle that has the motor built into the spindle. These spindles are generally used when higher spindle speeds (in excess of 16,000 rpm) are required. These spindles are more costly compared with the belt spindles.

No matter the type of spindle, the motor that drives the spindle is obviously important. Motors with two sets of windings, called Dual Wound Spindle Motors, provide more cutting torque and material removal. Single wound motors are used where lower torque is sufficient and higher base speeds are not an issue.

4. Enemies of the Spindle

The two major enemies of the spindle are: ① heat, ② contaminants (namely, chips and coolant invading the bearing system). Find out what design features are included (or available as options) that protect the spindle. Historically, the most common cause of spindle failure has been bearing failure due to contamination from coolant ingress, condensation, contamination or chip damage. If the operator wants the spindle temperature to stay cool, he should make sure contaminants stay out.

In most cases, contaminants enter the spindle because the spindle seal failed. Find out what design measures the machine tool manufacturer has taken to keep the seal tight. An air purge system uses a labyrinth seal and purges the seal with positive air pressure to keep contaminants out. A dual air purge system, a system with two ports (usually upper and lower) is one design feature that works well to keep contaminants out of the way.

Temperature is the other factor that leads to spindle problems. Because heat causes steel to expand, manufacturers should explain what measures they have taken to protect the spindle from head growth—which leads to mostly Y and Z axis changes.

Heat exchangers or chillers (most common) are used to keep the spindle cool and control spindle growth as well as head growth. This type of system adds life to the spindle and reduces head growth, and is typically used when you're running long cycles or high-duty cycles. The selection of the chiller is dependent upon the application. For extended high-speed applications, you may want to investigate a thermal stabilization system. This system uses a thermostat with an oil chiller to automatically cool the spindle as needed.

Another contributing factor to spindle performance is the used tools. Using unbalanced tools, worn tools and/or tools that are too long can affect the longevity of the spindle.

5. Cooling Considerations

Like the spindle, temperature can have a negative impact on tooling. Find out if the spindle comes with a coolant ring or uses flexible coolant nozzles. With a coolant ring, someone wants to find out how many nozzles there are and whether they are adjustable. Obviously, the more nozzles the better. Having the ability to adjust the direction of the nozzles is an advantage to cover a large range of tool lengths without frequent adjustments.

Coolant through the spindle (CTS) is generally recommended when machining at 12,000 rpm or more, and operators have custom tools or expensive tools that they want to ensure are protected. CTS are recommended at lower RPM for certain applications and duty cycles. Prices vary for this feature depending upon the pressure of the CTS and how the spindle was designed.

6. Replacement Costs

Just like tires on a car, drivers will eventually need to replace the spindle on their vertical machining center. In one's zeal to make a purchasing decision for a brand new machining center, make sure he looks ahead to the day when the spindle will need to be replaced. He need to know how much it will cost, availability of his type of spindle and downtime to install.

5.2.2 Guide Rail Design

The first frame subsystem design to consider would be a conventional railing system, which consists of a linear motion bearing and shaft assembly which would simply allow unrestricted movement along their lengths. The most logical rail design to consider, given the design specifications and size requirements, would be the sort of railing that could be supported in some way to handle the loads applied to it without much deflection. For instance, the railing system shown here has a simple steel shaft railing system and is lightweight. For many years there have been vast improvements made in rail design to help increase the performances of the rail system.

Steel shaft railing is both a simple and efficient design for linear motion applications. The shaft provides support to loading applications along the shaft, along with forces generated from linear motion, which makes this a perfect concept for this particular system.

A railing system uses a shaft and support system to support loading applications along the shaft, with the forces from linear motion. The shaft and support system in this particular system can come in a ceramic material which provides enhanced properties of the system. The enhanced properties include a reduction of vibration while also reducing deflection of the shaft during loading cases to help increase the life of the shaft.

The rail design is very diversely designed, and has been well engineered for loading applications. Each system found, even from several different vendors, has rail systems ranging from ceramic rails to case hardened steel railing systems. Most rail systems are case hardened steel, and have some sort of bearing to go along with them.

The V-notch rail system uses a notch in the rail and V-grooved wheel riding on the railing surface to carry the load and support linear motion. The V-notch rail can be more complex by notching the top and bottom of the rail which can be used for rails suspended above the ground, which makes this system a perfect concept for this particular system (Figure 5.8 and Figure 5.9).

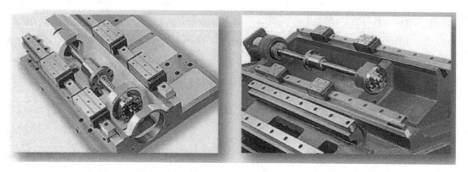

Figure 5.8 Rolling Guide

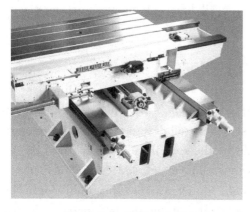

Figure 5.9 Sliding Guide

5.2.3 Rotated Working Table Design

A rotary table is a precision work positioning device used in metalworking. It enables the operator to drill or cut work at exact intervals around a fixed (usually horizontal or vertical) axis. Some rotary tables allow the use of index plates for indexing operations, and some can also be fitted with dividing plates that enable regular work positioning at divisions for which indexing plates are not available. A rotary fixture used in this fashion is more appropriately called a dividing head (indexing head).

The table shown in Figure 5.10 is a manually operated type. Powered tables under the control of CNC machines are now available, which provide a fourth axis to CNC milling machines. Rotary tables are made with a solid base, which has provision for clamping onto another table or fixture. The actual table is a precision-machined disc to which the work-

Mechanical Construction and Tool System of CNC Machines Chapter 5

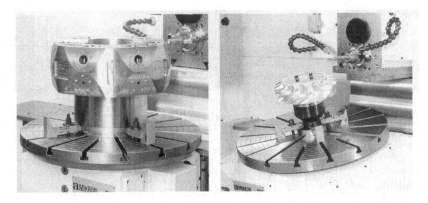

Figure 5.10 Rotated Working Table of CNC Machine

piece is clamped (T-slots are generally provided for this purpose). This disc can rotate freely, for indexing, or under the control of a worm (handwheel), with the worm wheel portion being made part of the actual table. High precision tables are driven by backlash compensating duplex worms.

A rotary table can be used:

(1) To machine spanner flats on a bolt.

(2) To drill equidistant holes on a circular flange.

(3) To cut a round piece with a protruding tang.

(4) To create large-diameter holes, via milling in a circular toolpath, on small milling machines that don't have the power to drive large twist drills ($>0.500"$ / $>13mm$).

(5) To mill helixes.

(6) To cut complex curves (with proper setup).

(7) To cut straight lines at any angle.

(8) To cut arcs.

(9) With the addition of acompound table on top of the rotary table, the user can move the center of rotation to anywhere on the part being cut. This enables an arc to be cut at any place on the part.

(10) To cut circular pieces.

Additionally, if converted to stepper motor operation, with a CNC milling machine and a tailstock, a rotary table allows many parts to be made on a mill that otherwise would require a lathe.

5.3 CNC Tool System

Automatic tool changer is shown in Figure 5.11. CNC machine tools can be used:

(1) For the parts having complex contours that cannot be manufactured by conventional machine tools.

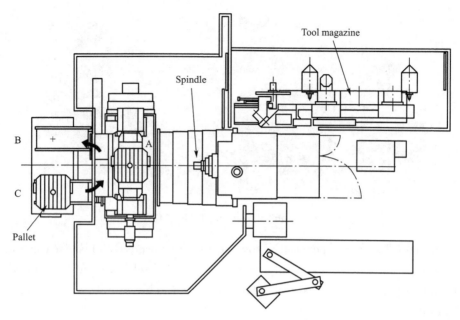

Figure 5.11 Typical Horizontal Machining Center With an Automatic Pallet Changer

(2) For small lot production, often for even single (one off) job production, such as for prototyping, tool manufacturing, etc.

(3) For jobs requiring very high accuracy and repeatability.

(4) For jobs requiring many setups and/or the setups very expensive.

(5) The parts that are subjected to frequent design changes and consequently require more expensive manufacturing methods.

(6) The inspection cost is a significant portion of the total manufacturing cost.

5.3.1 Cutting Tool

1. Cutting Tool Materials (Table 5.1)

(1) High speed steel.

(2) Cemented carbides.

Table 5.1 Summary of Applications for Various Cutting Tool Materials

Tool material	Work materials	Remarks
Carbon steels	Low strength, softer materials, non ferrous alloys, plastics	Low cutting speeds, low strength materials
Low/medium alloy steels	Low strength, softer materials, non ferrous alloys, plastics	Low cutting speeds, low strength materials
HSS	All materials of low and medium strength and hardness	Low to medium cutting speeds, low to medium strength materials

(续)

Tool material	Work materials	Remarks
Cemented carbides	All materials up to medium strength and hardness	Not suitable for low speed application
Coated carbides	Cast iron, alloy steels, stainless steels, super alloys	Not for titanium alloys, not for nonferrous alloys as the coated grades do not offer additional benefits over uncoated
Ceramics	Cast iron, Ni-base super alloys, non ferrous alloys, plastics	Not for low speed operation or interrupted cutting. Not for machining Al, Tialloys

The following guidelines would be useful for selecting a carbide grade:

(1) Choose a grade with the lowest cobalt content and the finest grain size consistent with adequate strength to eliminate chipping.

(2) Use straight WC grades if cratering, seizure or galling are not experienced in case of work materials other than steels.

(3) To reduce cratering and abrasive wear when machining steel, use grades containing TiC.

(4) For heavy cuts in steel where high temperature and high pressure deform the cutting edge plastically, use a multi carbide grade containing W-Ti-Ta and/or lower binder content.

(3) Coated carbides.

(4) Ceramics.

The following guidelines would be useful for selecting a ceramic cutting tool.

(1) Use the highest cutting speed recommended and preferably select square or round inserts with large nose radius.

(2) Use rigid machine with high spindle speeds and safe clamping angle.

(3) Machine rigid workpieces.

(4) Ensure adequate and uninterrupted power supply.

(5) Use negative rake angles so that less force is applied directly to the ceramic tip.

(6) The overhang of the tool holder should be kept to a minimum, not more than 1.5 times the shank thickness.

(7) Large nose radius and side cutting edge angle on the ceramic insert to reduce the tendency of chipping.

(8) Always take a deeper cut with a light feed rather than a light cut with heavy feed. Ceramic tips are capable of cuts as deep as one-half the width of the cutting surface on the insert.

(9) Avoid coolants with aluminum oxide based ceramics.

(10) Review machining sequence while converting to ceramics and if possible introduce chamfer or reduce feed rate at entry.

2. Cutter

A cutting tool (or cutter) is any tool that is used to remove material from the workpiece by means of shear deformation. Cutting may be accomplished by single-point or multipoint tools. Single-point tools are used in turning, shaping, planning, and similar operations, and remove material by means of one cutting edge. Milling and drilling tools are often multipoint tools. Grinding tools are also multipoint tools. Each grain of abrasive functions as a microscopic single-point cutting edge (although of high negative rake angle), and shears a tiny chip. The cutter structure is shown in Figure 5.12.

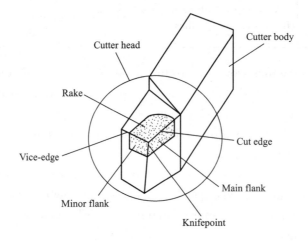

Figure 5.12　The Cutter

Cutting tools must be made of a material harder than the material which is to be cut, and the tool must be able to withstand the heat generated in the metal-cutting process. Also, the tool must have a specific geometry, with clearance angles designed so that the cutting edge can contact the workpiece without the rest of the tool dragging on the workpiece surface. The angle of the cutting face, the flute width, number of flutes or teeth, and margin size, are important. In order to have a long working life, all of the above must be optimized, plus the speeds and feeds at which the tool is run.

Standard insert shapes (Figure 5.13):

(1) V(35° diamond)—Used for profiling, weakest insert, 2 edges per side.

(2) D(55° diamond)—Somewhat stronger, used for profiling when the angle allows it, 2 edges per side.

(3) T(triangle)—Commonly used for turning because it has 3 edges per side.

(4) C(80° diamond)—Popular insert because the same holder can be used for turning and facing 2 edges per side.

(5) W(80° triangle)—Newest shape can turn and face like the C, but 3 edges per side.

(6) S(square)—Very strong, but mostly used for chamfering because it won't cut a square shoulder. 4 edges per side.

(7) R(round)——Strongest insert but least commonly used.

Mechanical Construction and Tool System of CNC Machines Chapter 5

R　　　　　S　　　　　W　　　　　C　　　　　T　　　　　D　　　　　V

Figure 5.13　Sandvik Coromant Shapes

Turning is a lathe operation in which the cutting tool removes metal from the outside diameter of a workpiece.

Figure 5.14　Turning

(a) External turning; (b) Internal turning; (c) Other turning

Tapping is the process of cutting a thread inside a hole so that a cap screw or bolt can be threaded into the hole. Also, it is used to make threads on nuts (Figure 5.15).

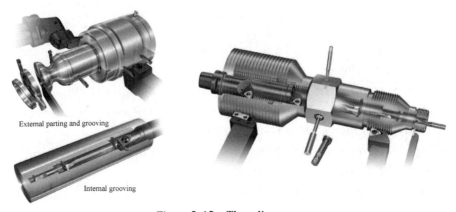

Figure 5.15　Threading

A vertical milling machine uses a rotating tool to produce flat surfaces, a very flexible, light-duty machine (Figure 5.16).

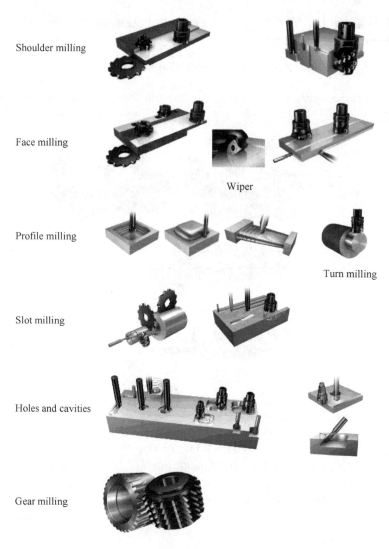

Figure 5.16 Milling

Boring is an operation to enlarge and finish holes accurately. This may be done on a lathe or a milling machine.

Grinding is an operation in which the cutting is done by the use of abrasive particles. Grinding processes remove very small chips in very large numbers by cutting the action of many small individual abrasive grains.

Drilling is an economical way of removing large amounts of metal to create semi-precision round hole or cavity.

Reaming is a sizing operation that removes a small amount of metal from a hole already drilled.

Honing is an internal cutting technique that uses abrasives on a rotating tool to produce extremely accurate holes that require a very smooth finish.

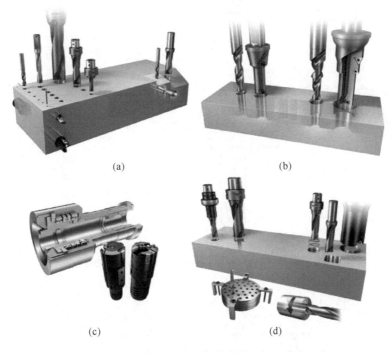

Figure 5. 17 Drilling

(a) General drilling; (b) Step and chamfer drilling; (c) Deep hole drilling; (d) Dedicated methods

3. Tooling Holding (Figure 5.18)

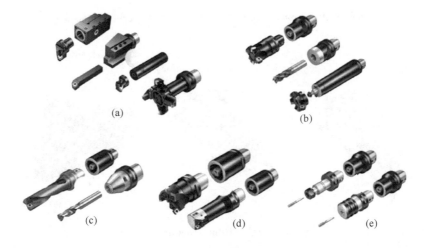

Figure 5. 18 Tool Holding

(a) Turning; (b) Milling; (c) Drilling; (d) Boring; (e) Tapping

Typical tool holding are shown in Figure 5.19, Figure 5.20 and Figure 5.21.

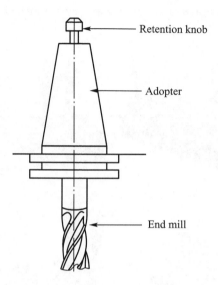

Figure 5.19　Typical Spindle Tool Holding an End Mill

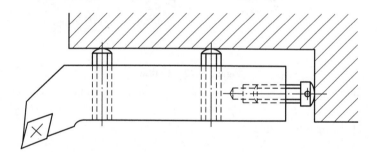

Figure 5.20　Typical Preset Tool Used in CNC Turning Machines

Figure 5.21　Milling Tool and Holder

4. Probe

These probes (Figure 5.22) can be used for a number of applications:

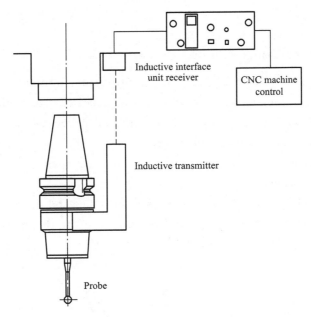

Figure 5.22 Inductive Transmission Systems Used in Machining Centers

(1) Datum of the workpiece.
(2) Workpiece dimension measurement.
(3) Tool offset measurement.
(4) Tool breakage monitoring.
(5) Digitizing.

The selection of CNC machine tool is shown in Figure 5.23.

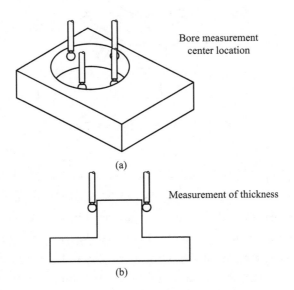

Figure 5.23 Probing Examples
(a) Inspection of a bore for diameter and center position; (b) Inspection of a web thickness

5.3.2　ATC

1. Automatic Tool Change Facilities

CNC tool machines are equipped with controllable automatic tool change facilities. Depending on the type and application area these tool change facilities can simultaneously take various quantities of tools and set the tool called by the CNC program into working position. The most common types are:

(1) The tool turret.

(2) The tool magazine.

The tool turret (Figure 5.24) is mostly used for lathes and the tool magazine for milling machines. If a new tool is called by the CNC program the turret rotates as long as the required tool achieves working position. Presently such a tool change only takes fractions of seconds.

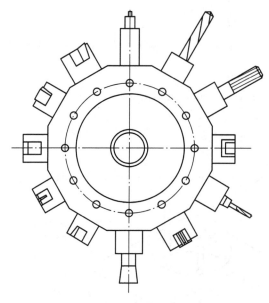

Figure 5.24　Example of a Turret

Depending

Depending on the type and size, the turrets of the CNC machines have 8 to 16 tool places. In large milling centers up to 3 turrets can be used simultaneously. If more than 48 tools are used tool magazines of different types are used in such machining centers allowing a charge of up to 100 and even more tools. There are longitudinal magazines, ring magazines, plate magazines, chain magazines (Figure 5.25) and cassette magazines.

In the tool magazine the tool change takes place using a gripping system also called tool changer (Figure 5.26). The change takes place with a double arm gripping device after a new tool has been called in the CNC program as follows:

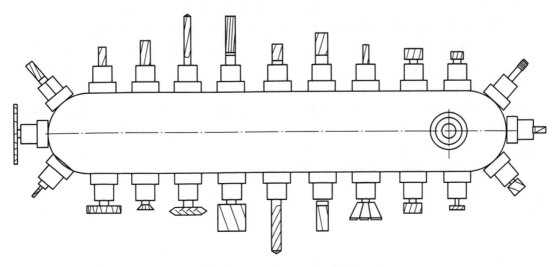

Figure 5.25　Example of a Chain Magazine

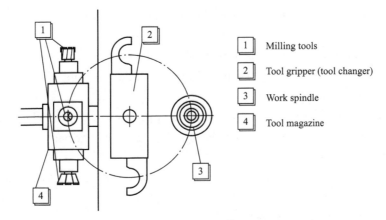

1	Milling tools
2	Tool gripper (tool changer)
3	Work spindle
4	Tool magazine

Figure 5.26　Automatic Tool Change Facility

(1) Positioning the desired tool in magazine into tool changing position.

(2) Taking the work spindle into changing position.

(3) Revolving the tool gripping device to the old tool in the spindle and to the new tool in the magazine.

(4) Taking the tools into the spindle and magazine and revolving the tool gripping device.

(5) Placing the tools into the spindle sleeve or magazine.

(6) Returning the tool gripping device into home position.

The tool change procedure takes between 6 to 15 seconds, whereby the quickest tool changers are able to make the tool change in merely one second.

2. ATC

For the ATC to operate, it is necessary to have the following:

(1) A tool magazine where sufficient number of tools can be stored.
(2) The tool adopter that has a provision for pick-up by the tool change arm.
(3) The ability in the control to perform the tool change function.
(4) Tool change procedure.

The tool change activity requires the following motions:

(1) Stopping the spindle at the correct orientation for the tool change arm to pick the tool from the spindle.
(2) Tool change arm moving to the spindle.
(3) Tool change arm picking the tool from the spindle.
(4) Tool change arm indexing to reach the tool magazine.
(5) Tool magazine indexing into the correct position where the tool from the spindle is to be placed.
(6) Placing the tool in the tool magazine.
(7) Indexing the tool magazine to bring the required tool to the tool change position.
(8) Tool change arm picking the tool from the tool magazine.
(9) Tool change arm indexing to reach the spindle X. New tool is placed in the spindle.
(10) Tool change arm moving into its parking position.

Security precautions on CNC machine tools.

The target of work security is to eliminate accidents and damages to persons, machines and facilities at work site.

Basically the same work security precautions apply to working on CNC machines as to conventional machine tools. They can be classified in three categories:

(1) Danger elimination.

① Defects on machines and on all devices necessary for work need to be registered at once.

② Emergency exits have to be kept free.

③ No sharp objects should be carried in clothing.

④ Watches and rings are to be taken off.

(2) Screening and marking risky areas.

① The security precautions and corresponding notifications are not allowed to be removed or inactivated.

② Moving and intersecting parts must be screened.

(3) Eliminating danger exposure.

① Protective clothing must be worn to protect from possible sparks and flashes.

② Protective glasses or protective shields must be worn to protect the eyes.

③ Damaged electrical cables are not allowed to be used.

Exercises

1. Write the type of drill from the box below under the correct picture (Figure 5.27).

   | high-helix drill | core drill | gun drill | oil hole drill |
   | spade drill | step drill | low-helix drill | straight-fluted drill |

   a. _____ b. _____ c. _____ d. _____

   e. _____ f. _____ g. _____ h. _____

   Figure 5.27 Exercise 1

2. Match each picture with the name of the lathe accessory (Figure 5.28).

   a. four-jaw chuck   b. steady rest   c. lathe center
   d. revolving tailstock center e. lathe dog   f. boring tools
   g. cam-lock spindle nose   h. tool holders   i. three-jaw chuck
   j. mandrel   k. follower rest   l. tool post

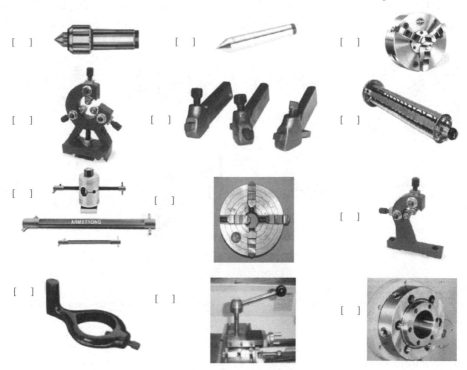

Figure 5.28 Exercise 2

3. Match the tool and accessory with each the picture (Figure 5.29).

a. fly cutter   b. end mills   c. woodruff key   d. boring head
e. vise   f. shell mill   g. keyseats   h. boring chuck
i. collet   j. slitting saw k. drill chuck   l. T-slot cutters

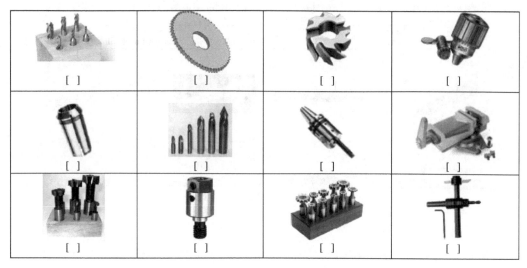

Figure 5.29 Exercise 3

4. Identify the following machining center accessories by writing their names under the picture (Figure 5.30).

automatic tool changer   oil-hole drill   two-flute end mills   single-point boring tool
rose reamer   face milling cutter   spiral flute tap   precision machine vise

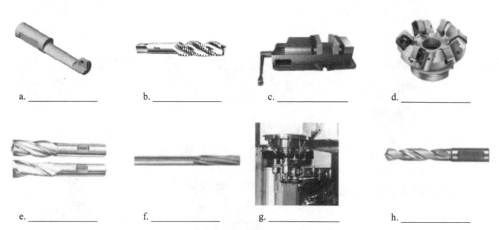

Figure 5.30 Exercise 4

5. A machining center accessory used for checking part dimensions is called____.
(a) A tool set
(b) A digital readout
(c) A trigger probe

(d) A dial indicator
6. Narrate the components of CNC Lathe.
7. What are the two major enemies of the spindle?
8. What are the functions of rotary table?
9. Narrate the main tool materials.

Chapter 6 Architecture for Modern CNC Technology

Objectives

- To understand the characteristics of open CNC system.
- To understand the advanced CNC technology application.

6.1 Open Architecture System for CNC Unit

After development by MIT in the early 1950s, CNC systems have advanced with the appearance and the advancement of the microprocessor. With the introduction of automation systems in the 1970s, the function of CNC system has made rapid progress. However, due to the complexity of NC technology, which requires not only fundamental control function but also various auxiliary technologies such as machining technology, process planning technology, and manufacturing technology, the market for NC system has been dominated by a few market leaders in Japan and Germany. The advanced manufacturers evolved CNC systems into closed systems in order to prevent their own technology from leaking out and keeping their market share. However, after the middle of the 1980s, a new manufacturing paradigm, where computer network and optimization techniques were applied to manufacturing systems with the progress of computer technology, has appeared together with the requirement for advanced control functions for high-speed and high-accuracy machining.

Closed CNC systems were not adequate for realizing the new manufacturing paradigm. The architecture of closed CNC systems could not meet the user's requirements and the improvement of CNC system was possible, not by MTB (Machine Tool Builders) but CNC system makers. The limited resources of CNC makers made it impossible to meet the new paradigm.

Therefore, various efforts to develop open CNC system have been made. As a typical result of these attempts, PC-NC was introduced in the early 1990s. Like IBM PC technology which appeared in the early 1980s and has progressed by third-party developments based on openness, CNC systems have progressed to PC-NC based on the openness of PC technology. However, now, despite low price, openness, and many developers of PC-NC, the lack of reliability and openness to application S/W has made it impossible to implement perfectly open systems (Figure 6.1).

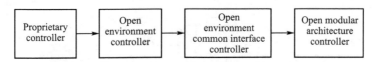

Figure 6.1 Progress of Open System Development

An open system is defined as a system that satisfies the following:

(1) Interoperability. This means the ability that the components compose the system cooperate to perform the specified task. For this ability, the standard specifications of the data representation language, behavior model, physical interface, communication mechanism, and interaction mechanism are needed. A bus-based system design is the most important.

(2) Portability. This means the ability for a component to be executed on the CNC system with different hardware or different software. Portability is very important from the commercial point of view. Since this means that a hardware device or a software module can be used on various platforms, it contributes to increasing the efficiency of a platform.

(3) Scalability. This means the ability to make extensions to or reductions of the system's functionality possible without large cost. Adding memory or a board to a PC is a typical example.

(4) Interchangeability. This means the ability to replace the existing component with a new component. Instead of replacing the whole system, replacing an existing motion board with a new algorithm is a typical example.

As the definition of an open system, modularity, extensibility, reusability, and compatibility can be considered. However, these can seem to belong to the above properties. From another point of view, an open system can be defined as a system with flexibility and standardization. Flexibility, though, has a similar meaning to interoperability and scalability, and standardization are similar to portability and interchangeability.

The presented scheme can be used as an economical upgrading method for small to medium industries. From the review of the researches done recently, the open architecture system is developed due to the fact of flexibility, improved of system delivery time and the quality assurance. In this research, an open architecture personal computer-based numerical control (OAPC-NC) system will be built based on these particular factors. Additionally, the OAPC-NC system should also allow the easy integration and reuse of hardware and software (Figure 6.2).

Open-system CNC products have changed rapidly. The higher speed communications choices available today have led to many different types of open architecture. Most of these open-system CNC systems integrate the "openness" of a standard PC with conventional CNC functions. The key advantage of specifying an open-system CNC is that it can allow the CNC features to remain current with the state of technology and the needs of the process even while the machine hardware ages. Among the capabilities that can be added to an open-system CNC via third-party software, some are more relevant and some are less relevant where mold machining is concerned. But across all shops using open-system CNC, some of the most common choices include:

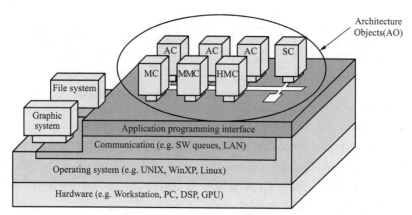

Figure 6.2　OAPC-NC Platforms with the Basic Application Objects

(1) Low-cost network communications.

(2) Ethernet.

(3) Adaptive control.

(4) Interfaces to bar code readers, tool ID readers and/or pallet ID systems.

(5) Mass part-program storage and editing.

(6) SPC (Statistical Process Control) data collection.

(7) Documentation control.

(8) CAD/CAM integration or shop-floor programming.

(9) Common operator interfaces.

The last item is particularly significant. A growing requirement is for the CNC to be easy to use. An important component of this ease of use is commonality of operation from CNC to CNC. Typically, operators must be trained separately for separating machines because the CNC interface differences between machine types and between machine tool builders. Open-system CNC systems provide new opportunities for working toward a control interface that's common throughout the shop.

Now, machine tool owners can design their own interfaces for CNC operation and they don't have to be one-C programmers to do so. In addition, open-system controls can permit individual log-on so personnel performing various functions: operator, programmer, maintenance and so on. See only the screens they need. Eliminating unnecessary screens makes CNC operation even more straightforward.

6.2　STEP-NC System

With the rapid advancement of information technology associated with NC technology, the manufacturing environment has changed significantly since the last decade. However, the low-level standard, G&M codes, have been used for over 50 years as the interface between CAM and CNC, and are now considered as an obstacle for global, collaborative and intelligent manufacturing. A new model of data transfer between CAD/CAM systems and

CNC machines, known as STEP-NC, is being developed worldwide to replace G&M codes. In this chapter, we will give an overview of STEP-NC and its related technology.

STEP-NC is expected to encompass the whole scope of E-manufacturing. The new STEP-NC data model has been developed for the replacement of the old standard G&M codes for milling, turning and EDM, and development on implementations is under way. Now the new data model has been established, development and implementation of STEP-compliant CAD/CAM/CNC systems based on the new data model are drawing worldwide attention.

STEP-NC is a new model of data transfer between CAD/CAM systems and CNC machines. As shown in Figure 6.3 (a), G-code contains just axis movement, spindle speed, feedrate, tool position in the tool changer and coolant. With this information, it is very difficult for machine operators to understand the operational flow, machining conditions and specification of tools only by reading a part program. Also, it is impossible for the CNC controller to execute an autonomous and intelligent control and to cope with emergency cases with this limited information. In contrast, STEP-NC contains the required functional information, as shown in Figure 6.3 (b), such as working step, machining feature, machining operation, machining tool, machining strategy, machine function and workpiece. In other words, STEP-NC includes a richer information set including "what-to-make" (geometry) and "how-to-make" (process plan).

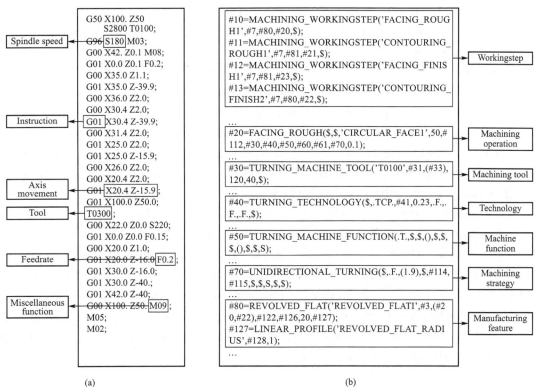

Figure 6.3 Comparison of G-code Part Program and STEP-NC Part Program
(a) G-code part program; (b) STEP-NC part program

6.3 Advanced Application of CNC Technology

6.3.1 Components of FMS

The flexible manufacturing cell (FMC) can be considered as a flexible manufacturing subsystem. The following differences exist between the FMC and the FMS (Flexible Manufacturing System):

(1) FMC is not under the direct control of the central computer. Instead, instructions from the central computer are passed to the cell controller.

(2) The cell is limited in the number of part families it can manufacture.

The following elements are normally found in a FMC: Cell controller, PLC, More than one machine tool, A materials handling device (robot or pallet).

The FMC executes fixed machining operations with parts flowing sequentially between operations.

A FMS consists of two subsystems: physical subsystem and control subsystem.

Physical subsystem of FMS includes the following elements:

(1) Workstations. It consists of CNC machines, machine-tools, inspection equipments, loading and unloading operation, and machining area.

(2) Storage-retrieval systems. It acts as a buffer during WIP (work-in-processes) and holds devices such as carousels used to store parts temporarily between work stations or operations.

(3) Material handling systems. It consists of power vehicles, conveyers, automated guided vehicles (AGV), and other systems to carry parts between workstations.

Control subsystem of FMS comprises of following elements:

(1) Control hardware. It consists of mini and micro computers, PLC, communication networks, switching devices and others peripheral devices such as printers and mass storage memory equipments to enhance the working capability of the FMS systems.

(2) Control software. It is a set of files and programs that are used to control the physical subsystems. The efficiency of FMS totally depends.

The programs which are created and simulated in the FMS programming system, for example, SL FMS (Figure 6.4) can be processed at the machine tool by means of the control module. The operation can be made alternatively by the keyboard and/or the mouse. A standard PC keyboard or a control-specific keyboard can be used as against the used numerical control. The modifications of the program are to be done in the respective editor of the appropriate programming system (Figure 6.5).

Architecture for Modern CNC Technology Chapter 6

Figure 6. 4   SL FMS

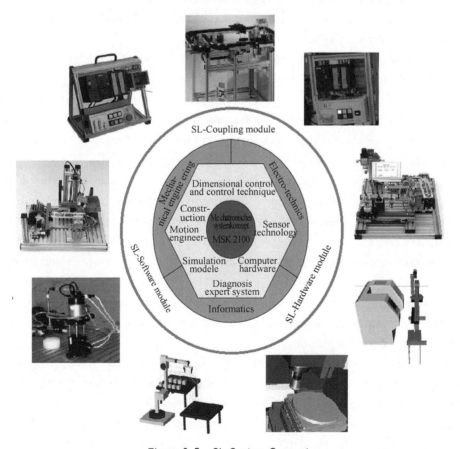

Figure 6. 5   SL System Concept

Basic features of the physical components of FMS are discussed below:

(1) Numerical control machine tools. Machine tools are considered to be the major building blocks of FMS as they determine the degree of flexibility and capabilities of the FMS. Some of the features of machine tools are described below.

① The majority of FMS use horizontal and vertical spindle machines. However, machining centers with vertical spindle machines have lesser flexibility than horizontal machining centers.

② Machining centers have numerical control on movements made in all directions, e. g., spindle movement in X, Y, and Z directions, rotation of tables, tilting of table, etc. to ensure the high flexibility.

③ The machining centers are able to perform a wide variety of operations e. g., turning, drilling, contouring etc. They consist of the pallet exchangers interfacing with material handling devices that carry the pallets within and between machining centers as well as automated storage and retrieval systems.

(2) Work holding and tooling considerations. It includes pallets/fixtures, tool changers, tool identification systems, coolant, and chip removal systems. It has the following features:

① Before machining is started on the parts, they are mounted on fixtures. So, fixtures must be designed in a way, to minimize part-handling time. Modular fixing has come up as an attractive method to fix a variety of parts quickly.

② The use of Automated Storage and Retrieval System (AS/RS) and material handling systems, such as AGV, lead to high usage of fixtures.

③ All the machining centers are well equipped with tool storage systems called tool magazines. Duplication of the most often used tools in the tool magazines is allowed to ensure the least non-operational time. Moreover, employment of quick tool changer, tool regrinding and provision of spares also help for the same.

(3) Material-handling equipments. The material-handling equipments used in flexible manufacturing systems include robots, conveyers, automated guided vehicle systems, monorails and other rail guided vehicles, and other specially designed vehicles. Their important features are:

① They are integrated with the machine centers and the storage and retrieval systems.

② For prismatic part material-handling systems are accompanied with modular pallet fixtures. For rotational parts industrial robots are used to load/unload the turning machine and to move parts between stations.

③ The handling system must be capable of being controlled directly by the computer system to direct to the various workstation, load/unload stations and storage area.

(4) Inspection equipment. It includes coordinate measuring machine (CMM) used for offline inspection and programmed to measure dimensions, concentricity, perpendicularity, and flatness of surfaces. The distinguishing feature of this equipment is that it is well integrated

with the machining centers.

(5) Other components. Other components include a central coolant and efficient chip separation system. Their features are:

① The system must be capable of recovering the coolant.

② The combination of parts, fixtures, and pallets must be cleaned properly to remove dirt and chips before operation and inspection.

6.3.2 CNC Technology for Mold Applications

CNC technology is changing rapidly, which helps to improve the productivity of machine tools used in the mold industry. Faster central processing units (CPUs) are at the heart of many CNC changes. However, the improvements go beyond just faster processing, and the speed itself touches on many different CNC advances. With so much that has changed in recent years, it's worthwhile to present a summary of the state of mold-making CNC technology today.

1. Block Processing Time (BPT) and Beyond

As CPU speeds have increased and CNC manufacturers have incorporated this speed into highly integrated CNC systems, there have been phenomenal changes that increase CNC performance. The faster, more responsive systems do more than faster process program blocking. In fact, a CNC system that can process part program blocks at a very high rate may perform as well as a system that processes data at a slower rate, because there are other potential bottlenecks downstream that the overall feature content of the CNC system also has to address.

Most mold shops today intuitively understand that high speed machining requires more than just BPT. In many ways, the analogy of a race car illustrates why this is so. Should the fastest car win the race? Even a casual observer of racing knows there is more to it than this.

First, the driver's knowledge of the race track is important. He has to know a sharp curve is coming so he can slow down just enough to take the curve safely and efficiently. CNC look-ahead performs a similar role in high-feed-rate mold machining, giving the CNC advanced knowledge of the sharp curves coming up.

Similarly, how quickly the driver reacts to what other drivers do, and other unpredictable effects, can be compared to the CNC's servo loop times, including position loop, velocity loop and current loop.

Consider also the smoothness of the driver's execution as he goes around the track. Skillful braking and accelerating have a significant impact on performance. Bell-type Acc/Dec in the CNC system gives similar smoothness to machine tool acceleration. Look-ahead helps here as well, because it allows many small Acc/Dec adjustments to replace an abrupt Acc/Dec change.

The analogy also applies in other ways. The power of the engine can be compared to the drives and motors. The weight of the car can be compared to the mass of the moving elements of the machine tool. The strength and rigidity of the car can be compared to the strength and rigidity of the machine. And the CNC's ability to maintain a specified path error can be related to how well the driver keeps the car on the track.

One other way the analogy relates to the state of CNC today is this: A car that isn't one of the very fastest may not need the most skilled driver. In the past, it was only high-end CNC system that could maintain high accuracy at high speeds. Today, mid-level and low-end CNC systems are so powerful that they may also do an acceptable job. The high-end CNC still offers the best available performance. But perhaps for the machine, the lower-level CNC will permit the same performance as a CNC at the top of the line. It used to be that the CNC was the limiting factor determining the maximum feed rate in mold machining, but today the limiting factor is the mechanics of the machine. A better CNC won't deliver more performance if the machine itself is already operating at its performance limit.

2. Features Inherent to the CNC System

Here are some of the CNC features fundamental to many mold machining processes today:

(1) NURBS Interpolation. This technology for interpolating along curves instead of dividing curves into short, straight line segments is still gaining popularity. Most of the CAM packages for die/mold applications today now have an option for outputting NURBS-formatted part program. At the same time, more powerful CNC system have allowed CNC manufacturers to add five-axis NURBS capability, as well as NURBS-related features that deliver improved surface finish, smoother motor performance, faster cutting rates and smaller part program size.

(2) Finer command unit. Most CNC systems issue motion and positioning commands to machine axes using a command unit of 1 micron or coarser. Taking advantage of the increase in processing power, some CNC systems today offer a command unit of 1 nanometer (0.000001 mm). This control increment is 1,000 times finer, providing for improved accuracy. It also provides for smoother motor performance, which can allow some machines to accelerate faster without increasing the shock to the machine.

(3) Bell-shaped Acc/Dec. Also called "Jerk Control" or "S-curve Acc/Dec", bell-shaped Acc/Dec allows a machine tool to accelerate faster than linear Acc/Dec. It also provides less position error than various Acc/Dec types including linear and exponential.

(4) Look-ahead. This is a widely used term, with many performance differences separating the way the feature works on low-end versus high-end controls. In general, look-ahead lets the CNC pre-process the program to ensure superior Acc/Dec control. The number of look-ahead blocks can range from two blocks to hundreds of blocks depending on the CNC. The number of blocks required depends on factors such as the minimum part

program execution time and the Acc/Dec time constant, but blocks of look ahead is probably the minimum acceptable value.

(5) Digital servo control. Digital servo technology has improved significantly, and most CNC manufacturers can now offer a digital servo solution. Advances include faster communications, serial connections between the drive and CNC, and faster and more numerous digital signal processors. These advances have combined to allow CNC to control the servo loops more tightly and thus control the machine better.

The technology helps in many ways:

(1) Increasing the sample speed of the current loop, combined with better current control, results in the motor heating up less. This not only extends motor life, but also means there is less heat transfer to the ballscrew and therefore improved accuracy. Increased sampling speed can also make a higher velocity loop gain possible, helping to increase the overall performance of the machine.

(2) Because many newer CNC systems offer a high-speed serial connection to the servo system, the CNC can now get a lot more information about motor and drive operation through this communication link. This has resulted in improved maintenance features.

(3) Serial position feedback permits higher accuracy at high feed rates. As CNC system got faster, the position feedback rate became a bottleneck in determining how fast a machine could move. Conventional feedback is carried by a signal type that limits speed according to the sample rate of the CNC and the electronics of the external encoder. Serial feedback eliminates this bottleneck, allowing fine position feedback resolution even at high speeds.

Linear motors. This technology has improved significantly in recent years in both performance and acceptance. Some of GE Fanuc's advances have resulted in machine tool linear motors with a maximum force of 15,500 N (Newton) and a maximum acceleration of 30G. Other advances have led to smaller size, lighter weight and more efficient cooling. All of these changes serve to enhance the benefits linear motors offer over rotary motors: benefits that include higher Acc/Dec rates, superior position control and higher stiffness, improved reliability, and inherent dynamic braking.

3. Five-axis Machining (Figure 6.6)

Five-axis machining is increasingly being applied to complex mold work. The technology can reduce the number of setups and/or machine tools required to produce a part, thereby minimizing work-in-process inventory and reducing total manufacturing time.

As CNC systems have become more powerful, CNC manufacturers have been able to add more five-axis features. Capabilities once found only in high-end controls are now available in mid-range products. Most of these features have to do with making five-axis machining easier to use for shops that have little five-axis experience. Today, accessible CNC technology can deliver all of these benefits to the five-axis machining process:

(1) Eliminate the need for qualified tooling.

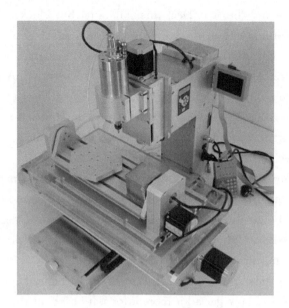

Figure 6.6 Five-Axis Machining

(2) Allow tool offsets to be set after the part program has been posted.

(3) Support "machine anywhere" programming, so that posted programs are interchangeable from machine to machine.

(4) Improve surface finish.

(5) Support various machine configurations, so the program no longer has to account for whether the spindle pivots or the workpiece pivots. This is now accounted for by parameters at the CNC.

One example of a five-axis machining feature specifically suited to mold machining is ball-nose end mill compensation. In order to properly compensate for a ball-nose end mill as the part or the tool pivots, the CNC must be able to dynamically adjust the cutter compensation vector in X, Y and Z. Better finish is one benefit of keeping the tool's contact point constant.

Other five-axis CNC functionality can be separated into the features related to pivoting the tool, features related to pivoting the part, and features that allow the operator to manually move the tool to a new vector.

When rotary axes pivot the tool, the tool length offset that normally affects only the Z axis now has components in X, Y and Z. In addition, tool diameter offsets that normally affect only the X and Y axes also have X, Y and Z components. And because the tool may be feeding in the rotary axes while it's cutting, all of these offsets have to be updated dynamically to account for continuous changes in the tool's orientation.

A CNC feature called "Tool Center Point Programming" can take care of this. The feature lets the programmer define the path and speed of the center point of the tool, while leaving it to the CNC to take care of the commands in the rotary and linear axes to ensure that the tool follows this programming. This feature makes the tool center point independent of the specific tool loaded into the machine.

(1) Tool offsets can be input at the machine tool just as in three-axis programming.

(2) Programs don't have to be re-posted to account for tool length changes.

The feature simplifies programming and posting for machines that achieve rotary-axis motion by pivoting the spindle.

Machines achieving rotary motion by pivoting the workpiece use similar functionality. Newer CNC system can compensate for this movement by dynamically adjusting fixture offsets and rotating coordinate axes to match the part's rotary motion.

The CNC can also have an important role when the operator is jogging the machine manually. Newer CNC systems allow the axis to be jogged in the direction of the tool vector. It is allowed that the tool vector to be changed without the location of the tool tip changing.

These features make a five-axis machine easier to use for 3+2 programming: the most common use of five-axis machines in mold making today. However, as new five-axis CNC features continue to evolve and gain acceptance, true five-axis mold machining is likely to become more common.

6.3.3 Virtual Axis Machine Tool

Virtual axis machine tool is named hexapod or parallel structured machine tool.

Hexapod, known as Stewart platform too, is a multi-axis machining center capable of full six degrees of freedom (DOF) motion plus spindle rotation at the tool head. Hexapod machines inherit all of the advantageous attributes of parallel mechanisms to enable more potential capabilities for manufacturing. Among these advantages, higher structural rigidity along with large payload capability and high speed motions will be capable for high speed and high accuracy machining.

Designed machine tool can be divided into the 6 basic features or subsystems:

(1) Main frame.
(2) Moving platform wit main milling spindle.
(3) Telescopic actuators.
(4) System of automatic tool changing.
(5) System of automatic part changing.
(6) Electrical switchbox with control and power part.

Figure 6.7 shows a basic structure of designed machine tool, the connection points, places for part changing by operator, moving platform with main milling spindle, electrical switchgear, etc. The main feature is a mechanism with parallel kinematic structure (PKS, hexapod) located in the middle of whole machine tool (Figure 6.8).

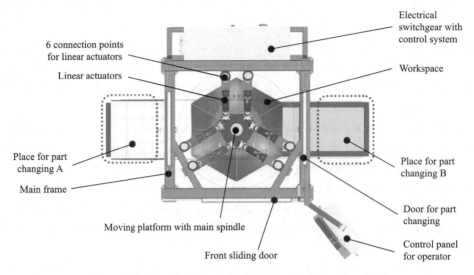

Figure 6.7　The Structure of a Machine Tool Based on PKS

Figure 6.8　Design of a Hexapod Machine Tool Without Cowling

6.3.4　Virtual CNC

The virtual CNC (VCNC) enables the prediction and optimization of a machine's dynamic performance at the design stage. By running part program on the virtual CNC and evaluating the contouring performance, the influence of various design choices such as guideway, drive, encoder, control law, and interpolation algorithm selection, can be assessed before the machine is actually built.

The VCNC can also be used for tuning servo control and interpolation parameters, without taking up production time on the actual machine tool. Once the desired response and contouring accuracy are achieved in the virtual model, the parameters can be directly implemented on the real machine with minimal down time. In process planning, the VCNC can be employed to evaluate the contour errors to different part programs, and make necessary changes to the feedrate and toolpath, to avoid tolerance violations due to servo errors. The simulation accuracy of VCNC relies on the utilization of realistic mathematical models to describe the dynamic behavior of each component.

Various applications have been developed, which take advantage of the VCNC's accurate simulation capability in predicting and improving the dynamic performance of real CNC machine tools. These are:

① Prediction of contour errors for part programming.
② Auto-tuning of servo controllers for feed drives.
③ Sharp corner tracking using spline interpolation.
④ Rapid identification of virtual drive models.

The architecture of the VCNC is shown in Figure 6.9, which resembles the real, reconfigurable and open CNC. The VCNC accepts reference toolpath commands generated on CAD/CAM systems in the form of industry standard Cutter Location (CL) format. The CL file is interpreted to realize the desired tool motion comprising of linear, circular, and spline segments. The axis trajectory commands are generated by imposing the desired feed profile on top of the toolpath commands. The feed profiling can be configured to employ piecewise constant, trapezoidal, or cubic acceleration transients.

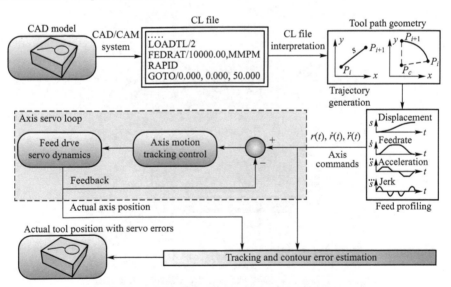

Figure 6.9 Architecture of Virtual CNC System

The axis servo loops are closed by configuring the motion control, feed drive, and feedback modules. The motion controller can be selected from a library of frequently used control laws such as P, PI, PID, P-PI cascade, and lead lag control, as well as more elaborate techniques proposed in literature like pole placement, generalized predictive, adaptive sliding mode, and feed forward control, as well as friction compensation.

The feed drive module can be configured to emulate the dynamics of direct or geared drives. Characteristics of the amplifier, motor, axis inertia, friction, and drive mechanism can be fully defined, including nonlinear effects such as quantization, current and voltage saturations, stick-slip friction, and axis backlash. Experimentally identified or analytically predicted high order drive models with structural resonances can also be incorporated. The feedback module can be configured from a combination of linear or angular position, velocity, and acceleration sensors, each with its user defined accuracy and noise characteristics. When the VCNC is assembled, its performance can be assessed by running various part programs and evaluating the servo tracking and contour errors, as well as axis velocity, acceleration, and jerk profiles, and motor torque and power histories. It is also possible to conduct frequency and time domain analyses, which aid the user in evaluating and improving the stability margins and servo performance of the VCNC axes.

6.3.5 3D Printer

A 3D printer cannot make any object on demand like the "Star Trek" replicators of science fiction. But a growing array of 3D printing machines (Figure 6.10) has already begun to revolutionize the business of making things in the real world.

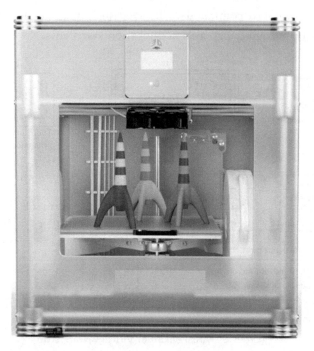

Figure 6.10　3D Printer Machine

3D printers work by following a computer's digital instructions to "print" an object using materials such as plastic, ceramics, and metal. The printing process involves building up an object one layer at a time until it's complete. For instance, some 3D printers squirt out a stream of heated, semi-liquid plastic that solidifies as the printer's head moves around to create the outline of each layer within the object (Figure 6.11).

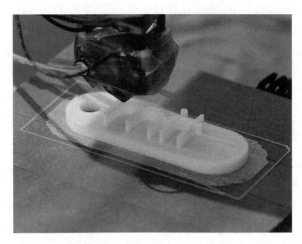

Figure 6.11 3D Printer Sample

The instructions used by 3D printers often take the form of computer-aided design (CAD) files, digital blueprints for making different objects. It means a person can design an object on their computer using 3D modeling software, hook the computer up to a 3D printer, and the 3D printer build the object right before his or her eyes (Figure 6.12).

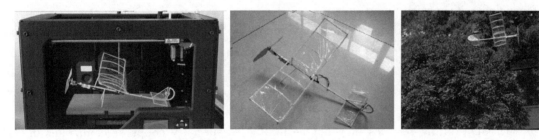

Figure 6.12 3D Printer Plane (Tongji University, China)

1. History of 3D Printing

Manufacturers have quietly used 3D printing technology, also known as additive manufacturing, to build models and prototypes of products over the past 20 years. Charles Hull invented the first commercial 3D printer and offered it for sale through his company 3D systems in 1986. Hull's machine used stereolithography, a technique that relies upon a laser to solidify an ultraviolet-sensitive polymer material wherever the ultraviolet laser touches.

The technology remained relatively unknown to the greater public until the second decade of the 21st century. A combination of the U.S. government funding and commercial startups has created a new wave of unprecedented popularity around the idea of 3D printing since that time.

First, American president Barack Obama's administration awarded US \$30 million to create the National Additive Manufacturing Innovation Institute (NAMII) in 2012 as a way of helping to revitalize U.S. manufacturing. NAMII acted as an umbrella organization for a network of universities and companies that aimed to refine 3D printing technology for rapid deployment in the manufacturing sector.

Second, a new wave of startups has made the idea of 3D printing popular within the so-called "Maker" movement that emphasizes do-it-yourself (DIY) projects. Many of those companies offer 3D printing services or sell relatively cheap 3D printers that can cost just hundreds rather than thousands of dollars.

2. Future of 3D Printing

3D printing probably won't replace many of the usual assembly-line methods for building standard products. Instead, the technology offers the advantage of making individual, specifically tailored parts on demand, something more suited to creating specialized parts for U.S. military aircraft rather than making thousands of trash cans for sale at WalMart. Boeing has already used 3D printing to make more than 22,000 parts used on civilian and military aircraft flying today.

The medical industry has also taken advantage of 3D printing's ability to make unique objects that might otherwise be tough to build using traditional methods. U.S. surgeons implanted a 3D-printed skull piece to replace 75% of a patient's skull during an operation in March 2013. Researchers also built a 3D-printed ear mold that served as the framework for a bioengineered ear with living cells.

The spread of 3D printing technology around the world could also shrink geographical distances for both homeowners and businesses. Online marketplaces already allow individuals to upload 3D-printable designs for objects and sell them anywhere in the world. Rather than pay hefty shipping fees and import taxes, sellers can simply arrange for a sold product to be printed at whatever 3D printing facility is closest to the buyer.

Such 3D printing services may not be limited to specialty shops or companies in the near future. Staples stores planed to offer 3D printing services in the Netherlands and Belgium starting in 2013.

Businesses won't be alone in benefiting from 3D printing's print-on-demand-anywhere capability. The U.S. military has deployed 3D printing labs to Afghanistan as a way to speed up the pace of battlefield innovation and rapidly build whatever soldiers might need onsite. NASA (National Aeronautics and Space Administration) has looked into 3D printing for making replacement parts aboard the International Space Station and building spacecraft in orbit.

Most 3D printers don't go beyond the size of household appliances such as refrigerators, but 3D printing could even scale up in size to build objects as big as a house. A separate NASA project has investigated the possibility of building lunar bases for future astronauts by using moon "dirt" known as regolith.

3. Limitations of 3D Printing

3D printing still has its limits. Most 3D printers can only print objects using a specific type of material, a serious limitation that prevents 3D printers from creating complex objects, such as an Apple iPhone. Yet researchers and commercial companies have begun developing workarounds. Optomec, a company based in Albuquerque, New Mexico, has already made a 3D printer capable of printing electronic circuitry onto objects.

The 3D printing boom could eventually prove disruptive in both a positive and negative sense. For instance, the ability to easily share digital blueprints online and print out the objects at home has proven a huge boon for DIY makers.

But security experts worry about 3D printing's ability to magnify the effects of digital piracy and the sharing of knowledge that could prove dangerous in the wrong hands. A Texas group has already begun pushing societal boundaries by working on the world's first fully 3D-printable gun.

Exercises

1. Narrate open architecture system of CNC.
2. What are the components of FMS? What is the function of each component?
3. What are the types of FMS? What is the feature of each type?
4. What is concept of 3D printer?

Appendix I Glossary

A

Abrasive
The material used in making grinding wheels or abrasive cloth.

Abrasive cut-off saw
A cutting-off machine used to cut material by means of a thin, abrasive wheel revolving at high speed.

Absolute system
A numerical control system in which all positional dimensions, both input and feedback, are given with reference to a common datum point.

Accident
Something going wrong unexpectedly, such as falling on a slippery floor.

Accountability
An obligation or willingness to accept responsibility to somebody or for something.

Accuracy
The ability to be precise and avoid errors.

Adjust
To bring the components of a part into a more effective, or efficient calibration, or state.

Adjustable center rest
Part of the universal dividing head set that prevents long, slender work held between centers from bending.

Air gauging
A technology that employs the use of air flow volumes and air pressure to determine the size of measured part dimensions.

Alpha iron
The state in which iron exists below the lower critical temperature. In this state, the atoms form a body-centered cube.

Angle
The space between two lines diverging from a common point and expressed in numerical form.

Angle of keenness
The included angle produced by grinding side rake and side clearance on a toolbit.

Angular cutting
A bandsaw operation where the work may be clamped at any angle and fed through

the saw.

Annealing
A heat-treating operation used to soften metal and to improve its machinability.

Apron
Part of the lathe that is fastened to the saddle and houses the gears and mechanism required to move the carriage and cross-slide automatically.

Arbor support
The part of the milling machine that aligns and supports various arbors and attachments.

Assembly
The process of joining together two or more parts to complete a unit or structure.

Austenite
A solid solution of carbon in iron, which exists between the lower and upper critical temperatures.

Automatic tool change (ATC)
One of the CNC components of a machining center.

B

Back rake angle
The backward slope of the tool face away from the nose.

Base
(1) The bottom surface of the tool shank.
(2) The part of a vertical mill that is made of ribbed cast iron and may contain a coolant reservoir.

Bed
Part of the lathe usually made of cast iron that provides a heavy rigid frame on which all the main components are mounted.

Bench vise
A device for holding small work securely for sawing, chipping, filing, polishing, drilling, reaming, and tapping operations.

Bevel protractor
An instrument in the form of a half circle used for measuring and making angles other than a 90° angle.

Blade tension handle
The part of the horizontal band saw used to adjust the tension on the saw blade.

Blueprint
A generic term for the representation of the work piece to be machined. It can be a pencil sketch or a CAD drawing or any of a number of other graphic options used to represent the design.

Body

The portion of the drill between the shank and the point.

Bond

The media or glue that holds abrasive grains together in the form of a wheel.

Boring

The operation of truing and enlarging a hole by means of a single point cutting tool.

Boring head

A cutting tool placed in a cutter head to dress round holes.

Brightness

The intensity of light reflected or emitted by something.

Brittleness

The property of the metal that permits no permanent distortion before breaking.

Broaching

A process in which a special tapered multi-toothed cutter is forced through an opening or along the outside of a piece of work to enlarge or change the shape of the hole or to form the outside to a desired shape.

Built-up edge

A layer of compressed metal from the material being cut which adheres to and piles up on the face of the cutting tool edge during a machining operation.

Burnishing

A process that develops a smooth finish on a metal by tumbling or rubbing with a polished hand tool.

Bushing

An operation in the manufacture of vitrified grinding wheels in which the arbor hole is fitted with a lead or plastic-type liner to fit a specific spindle size.

C

Careless

Not giving sufficient attention or thought to avoiding harm or mistakes.

Carriage

Part of the lathe that consists of three main parts: the saddle, cross slide, and apron. It is used for mounting and moving most of the cutting tools along the lathe bed.

Cementite

A carbide of iron, which is the hardener of steel.

Center head

A part of the combination square set that is used to find the center of or to bisect a round or square workpiece.

Chip conveyor

A moveable belt that helps to remove chips from the machine.

Chip-tool interface
That portion of the face of a cutting tool on which the chip slides as it is cut from the metal.

Chuck
A 3-jaw (self centering) or 4-jaw (independent) device used to clamp the part being machined on a lathe.

CMM
Coordinate Measuring Machine. It is a mechanical system designed to move a measuring probe to determine coordinates of points on a workpiece surface.

CNC code
Contains a program of instructions and commands, which consists of numbers, letters, and other symbols.

Cold circular cutoff saw
A cutting-off machine used to cut soft or unhardened metals.

Collet
A cone-shaped sleeve used for holding circular or rod-like pieces in a lathe or other machines.

Column
(1) An air-electronic amplifier or a flow system amplifier featuring a vertical bar graph display or flowmeter tube.

(2) The part of a vertical mill that holds the turret.

Column face
A precision-machined and scraped section of the horizontal milling machine used to support and guide the knee when it is moved vertically.

Combination set
A set of tools used extensively in layout work. It consists of a steel rule, square head, bevel protractor, and center head, compound rest part of the lathe that is mounted to the cross slide and used to support the cutting tool.

Computer-aided design (CAD)
The use of a wide range of computer-based tools that assist engineers, architects, and other design professionals in their design activities.

Computer-aided manufacturing (CAM)
Refers to the software used to generate the instruction codes for a CNC machine in order to cut out a shape designed in a computer-aided design (CAD) system.

Computer Numerical Control (CNC)
A form of programmable automation in which the machine tool is controlled by a program in computer memory.

Continuous path positioning
A CNC programming method that has the ability to control motions on two or more machine axes simultaneously to keep a constant cutter-work relationship.

Contour

A curved surface or dimension that is cut into a workpiece.

Coolant

A substance that is used to prevent a workpiece from reaching excessively high temperatures during machining.

Counterboring

The operation of enlarging the top of a previously drilled hole to a given depth to provide a square shoulder for the head of a bolt or cap screw.

Countersinking

The operation of producing a tapered or cone-shaped enlargement to the end of a hole.

Crest

The top surface joining two sides of a thread.

Crossfeed handwheel

Part of the horizontal milling machine used to move the table toward or away from the column.

Cross slide

Part of the lathe mounted on the traverse slide of the carriage, and uses a handwheel to feed tools into the workpiece.

Cross traverse hand wheel

Part of the vertical milling machine that moves the work table in and out.

Crystal elongation

The distortion of the crystal structure of the work material that occurs during a machining operation.

Cutting edge

The leading edge of the tool bit that does the cutting.

Cutting tools

Any tools used to cut material. This usually refers to milling cutters, lathe bits, or drills.

D

Damage

Loss or harm resulting from injury to person, or property.

Datum

A reference point from which movements or measurements are made. A datum can lie anywhere on the surface of a workpiece.

Deformed zone

The area in which the work material is deformed during cutting.

Density

The mass per unit volume of a substance, usually expressed in grams per cubic centimeter or in pounds per cubic foot.

Glossary Appendix I

Depth

The distance between the crest and root of a thread measured perpendicular to the axis.

Depth micrometer

An instrument used to measure the depth of slots, steps, and other features.

Diagnostic

Information that is vital to routine machine maintenance.

Die

A tool to make threads on the outside diameter of shafts. It is also used to form and blank sheet metal parts.

Dismantling

Taking apart, disassembling or tearing down.

Dividers

Instruments used for scribing arcs and circles on a layout and for transferring measurements.

Dividing head

An attachment for the milling machine that divides the workpiece circumference accurately into any number of divisions.

Divisor

A number divided into another number. For example, 7 is a divisor of 42 because $42/7=6$.

Drill

A tool used to make holes in firm material.

Drill chuck

A holding device used to hold and drive straight-shank cutting tools.

Drilling

The operation of producing a hole by removing metal from a solid mass using a cutting tool called a twist drill.

Ductility

The ability of the material to be permanently deformed without breaking.

E

Elasticity

The ability of the material to return to its original shape.

Electropolishing

A process that involves passage of electric current through a workpiece, while it is submerged in a specially-designed acid solution.

Elevating screw

The part of the horizontal milling machine that gives an upward or downward movement to the knee and the table.

Emergency stop
Used for emergencies only, the control button that automatically shuts down all machine functions.

End cutting edge angle
The angle formed by the end cutting edge and a line at right angles to the centerline of the tool bit.

End mill
A milling cutter with straight or tapered shanks that can cut both on the sides and on the end.

End relief angle
The angle ground below the nose of the tool bit which permits the cutting tool to be fed into the work.

F

Face
The surface against which the chip bears as it is separated from the work.

Feed
The motion of moving the workpiece and the cutting tool together so as to remove material.

Feed dial
The part of the horizontal milling machine used to regulate the table feeds.

Feed reverse lever
Part of the lathe mounted on the headstock that reverses the rotation of the feed rod and lead screw.

Feed rod
Part of the lathe that advances the carriage for turning operations when the automatic feed lever is engaged.

File
A hand cutting tool made of high-carbon steel. It has a series of teeth cut on its body by parallel chisel cuts.

File card
A wire brush mounted on a block of wood to clean the file.

Finish
Refers to the surface appearance of steel after final treatment.

Finished product
The goods or services produced and completed by a company.

Firing
An operation in the manufacture of vitrified grinding wheels which causes the bond to melt and form a glassy case around each grain producing a hard wheel.

Fitting
A mechanical device used to attach two pieces of tubing/piping together or to attach a piece of tubing/pipe to a component.

Fixture
A production work-holding device used for machining duplicate workpieces.

Flank
The surface of the tool adjacent to and below the cutting edge.

Flowmeter tube
A graduated glass tube of a precise size with a "floating" cork that displays the reading on a flow air gauge system.

Fly cutter
A unit holding two or more replaceable cutting tools that is used for milling large, flat surfaces.

Follow rest
Part of the lathe bolted to the carriage that uses adjustable fingers to bear against the workpiece opposite the cutting tool to prevent deflection.

Form turning
A lathe operation that forms irregular shapes or contours on a workpiece.

Fraction
A number that is not a whole number, e.g., 1/2 simple fraction or 0.5 decimal fraction, formed by dividing one quantity into another.

Frame
A part of the horizontal band saw, hinged at the motor end, which has two pulley wheels mounted on it, over which the continuous blade passes.

Friable
Easily crumbling a solid into powder or small particles.

Friction sawing
A burning process by which a saw band, with or without saw teeth, is run at high speeds to burn or melt its way through the metal.

Full scale value (FSV)
The numeric equivalent of the graduated display. It is usually 1.5 to 2 times greater than the tolerance being measured to show approach or oversize conditions.

G

Gage
The thickness of the saw blade that has been standardized according to blade width.

Gamma iron
The state in which iron exists in the critical range.

Gauge
Any one of a large variety of devices for measuring or checking the dimensions of ob-

jects.

G-codes
Refer to some action occurring on the X, Y, and/or Z axis of a machine tool that cause some movement of the machine table or head.

Gearbox
Part of the lathe inside the headstock, providing multiple speeds with a geometric ratio by moving levers.

Grade
The degree of strength with which the bond holds the abrasive particles in the bond setting on a grinding wheel.

Grinding wheel
An expendable wheel composed of abrasive material held together with a suitable bond and used in grinding machines.

Grooming
To care for one's appearance such as wearing approved safety clothes.

Grooving
A lathe operation done at the end of a thread to permit full travel of the nut up to a shoulder or at the edge of a shoulder to ensure a proper fit of mating parts.

H

Hammer
A hand tool consisting of a shaft with a metal head at right angles to it, used mainly for driving in nails and beating metal.

Hand hacksaw
A tool consisting of a frame and a saw blade generally used for cutting metal into pieces.

Handle
A rotating knob on the CNC control panel that moves the machine components along the axes. The handle "clicks" in controlled, measured increments that an operator selects on the keypad.

Hand reamer
A tool used to finish drilled holes accurately and provide a good finish.

Hardening
The process of heating steel above its lower critical temperature and quenching in the proper medium of water, oil, or air to produce martensite.

Hardness
The ability of the material to resist penetration.

Hazard
A danger or harm.

Head
That part of the drive system on the vertical milling machine that transforms electrical power from a motor to mechanical power in the spindle.

Headstock
Part of the lathe mounted in a fixed position on the inner ways, usually at the left end. Using a chuck, it rotates the work.

Heat treatment
The process of heating and subsequent cooling of metals to produce the desired mechanical properties.

Height gauge
An instrument used to scribe accurately dimensioned lines on a workpiece which has been prepared by brushing it with layout dye.

Hermaphrodite caliper
A tool for marking lines parallel to square edges and shoulders on a workpiece.

Honing
The process of removing stock generally on the internal cylindrical surface of a workpiece with an abrasive stick mounted in a holder.

horizontal band saw
A cutting-off machine that has a flexible, belt-like "one-way" blade that cuts continuously in one direction.

Housekeeping
Management and maintenance of the property and equipment of the shop.

I

Impedance
A measure of resistance to electrical current flow when a voltage is moved across something, such as a resistor.

Incremental system
A CNC programming mode in which program dimensions or positions are given from the current point.

Index crank
A device consisting of an arm or handle that is connected to an index head used to turn the spindle.

Indexing
The process of providing discrete spaces, parts, or angles in a workpiece by using a dividing head.

Index plates
Plates with circular graduations or holes arranged in circles, each circle with different spacing used for indexing on machines.

Injury

Damage or harm.

Installing

Putting machinery or equipment into place and making it ready for use.

ISO

The abbreviation for the International Organization for Standardization.

J

Jig

The part of a metalworking machine that holds the object to be worked on and guides the cutting or drilling tool.

K

Keyseat

A cutter that makes a recessed groove or slot machined into a shaft or part going on the shaft, usually a wheel or gear.

Keyway

A mechanical locking device located on the slitter head spindle shaft that holds the knives and spacers in place.

Knee

(1) The part of the vertical milling machine that moves up and down by sliding on ways that are parallel to the column and supports the saddle and table.

(2) The part of the horizontal milling machine that houses the feed mechanism.

Knurling

A late operation that impresses a diamond-shaped or straight-line pattern into the surface of the workpiece to improve its appearance or to provide a better gripping surface.

L

Ladder diagram

Ladder diagrams are specialized schematics commonly used to document industrial control logic systems. They are called "Ladder" diagrams because they resemble a ladder, with two vertical rails (supply power) and as many "rungs" (horizontal lines) as there are control circuits to represent.

Lapping

An abrading process used to remove minute amounts of metal from a surface that must be flat, accurate to size, and smooth.

Lathe

A turning machine capable of producing round diameters by rotating a workpiece against a stationary single-point cutting tool.

Layout
The process of measuring and marking a workpiece with finely scribed lines that guide you during the cutting process.

Layout die
A fast-drying colored liquid, usually a very deep blue, that is brushed onto a workpiece to prepare it for laying out guide lines.

Lead
The distance a screw thread advances axially in one revolution.

Lead screw
Part of the lathe used for cutting threads.

Lobe
A rounded projection that is part of a larger structure.

Longitudinal traverse hand wheel
Part of the vertical milling machine that moves the worktable to the left and right.

Loop
An instruction or series of instructions that is to be executed repeatedly to produce a desired operation.

M

Machine control unit (MCU)
A small, powerful computer that controls and operates a CNC machine.

Machine tool
An apparatus consisting of inter-related parts with separate functions, used to remove material from a workpiece. Examples are milling machines, lathes, drill presses, surface grinders, and hundreds more.

Machining
To cut, shape, or finish a piece of work using a power-driven tool such as a lathe or drilling device.

Machining center
A CNC machine which at one setup is capable of executing such operations as milling, drilling, boring, tapping, reaming and so forth, on one or more faces of the part.

Macro instruction
A statement, typically for an assembler, that invokes a macro definition to generate a sequence of instructions or other outputs.

Magnification
The visual increase of size that is created by an air amplifier.

Malleability
The ability of the material to be rolled into shapes.

Martensite
The structure of fully hardened steel obtained when austenite is quenched.

Mating part

Either one of a pair of things that belong together.

M-codes

Used to turn either ON or OFF different functions that control certain machine tool operations.

MDI/DNC

An operation mode key that lets an operator either enter and execute program data without disturbing stored data when pressed once, or execute programs from a centrally located computer storage device when pressed twice.

Measuring instruments

Devices used to determine physical size of parts to verify compliance to requirements.

Micrometer

A precision measuring instrument, used by machinists. Each revolution of the rachet moves the spindle face 0.5mm towards the anvil face.

Milling machine

Used to produce flat and angular surfaces, grooves, contours, gears, racks, sprockets, and helical grooves.

Mold

A hollow form for giving a particular shape to something in a molten or plastic state.

Molding

An operation in the manufacture of vitrified grinding wheels in which a proper amount of the mixture is placed in a steel mold of the desired wheel shape and compressed in a hydraulic press.

Motor

The part of the vertical milling machine mounted on top of the head and providing drive to the spindle, usually through V-belts.

Multiplier

The number by which another number multiplicand is multiplied, e.g. the number 4 is the multiplier in the statement $2 \times 4 = 8$.

N

Normalizing

The process of heating metal to just above its upper critical temperature and cooling it in still air to remove internal stresses and strains and to improve machinability.

Nose

The tip of the cutting tool formed by the junction of the cutting edge and the front face.

Nose radius

The radius to which the nose is ground.

Notching
A bandsaw application where sections of metal can be removed in one piece rather than in chips.

Nozzle
The orifice in the air gauge tooling that emits the air that blows against the part being measured.

Numerical Control(NC)
A form of programmable automation in which the machine tool is controlled by punched tape.

O

Offset
A numerical value stored in the CNC control that repositions machine components. Offsets are used to adjust for differences in tool geometry, part size, tool wear, etc.

Overarm (ram)
(1) Part of the vertical milling machine that slides on the turret and allows the milling head to be repositioned over the table.

(2) Part of the horizontal milling machine that provides for correct alignment and support of the arbor and various attachments.

P

Pallet
An automatic moveable table that supports a workpiece and slides or pivots into and out of the machining center. Multiple pallets allow an operator to set up a part while another is being machined.

Parallel
Two lines or surfaces extending in the same direction, everywhere equidistant, and not meeting.

Part program
A series of instructions used by a CNC machine to perform the necessary sequence of operations to machine a specific workpiece.

PCBN
An acronym for Polycrystalline Cubic Boron Nitride, which is a crystalline body of many small crystals randomly oriented to form a material for cutting hard ferrous metals.

Pearlite fine
A laminated structure of ferrite, usually the condition of steel before heat treatment.

Pitch
(1) The number of teeth per linear inch of the bandsaw blade.

(2) The distance from a point on one thread to a corresponding point on the next thread, measured parallel to the axis.

Plastic deformation

The change in shape of the work material that occurs in the shear zone during a cutting action.

Plastic flow

The flow of metal that occurs on the shear plane, which extends from the cutting-tool edge to the corner between the chip and the work surface.

Pliers

A hand tool with two hinged arms ending in jaws that are closed by hand pressure to grip something.

Point

(1) The end of the tool that has been ground for cutting purposes.

(2) The part of a twist drill that consists of the chisel edge, lips, lip clearance, and heel.

Point-to-point positioning

A CNC programming method in which any number of programmed points are joined together by straight lines.

Power hacksaw

A cutting-off machine utilized to cut material of various shapes and sizes up to six inches across.

Prevention

Taking advance measures against something possible or probable such as measures taken to prevent leaks.

Prick punch

Instrument used to permanently mark the location of layout lines.

Punch press

A power driven machine used to cut, draw, or otherwise shape material, especially metal sheets, with dies, under pressure or by heavy blows.

Q

Quadrant

A machine part that is shaped like a quarter circle.

Quick change gearbox

Part of the lathe that contains a number of different-size gears and provides the feed rod and lead screw with various speeds for turning and thread cutting operations.

Quill

Part of the vertical milling machine that moves vertically in the head and contains the spindle.

Quill feed hand wheel

Part of the vertical milling machine that moves the quill up and down within the head as does the quill feed lever.

R

Radius cutting

A bandsaw operation where internal or external contours may be cut easily.

Reaming

The operation of sizing and producing a smooth, round hole from a previously drilled or bored hole with the use of a cutting tool having several cutting edges.

Reference

A surface of known flatness or a point from which other lines and locations can be measured.

Refining

A process that produces liquid steel, suitable for ladle metallurgy treatment.

Reset

Returning a storage location in the MCU to zero or to a specified initial value.

Rigging

Gates, risers, loose pieces, etc., needed on the pattern to produce a sound casting.

Roller guide brackets

A part of the horizontal band saw that provide rigidity for a section of the blade and can be adjusted to accommodate various widths of material.

Root

The bottom surface joining the sides of two adjacent threads.

Route sheet

A document that describes the order of processing for the part(s) being manufactured (machined).

Rupture

The tear that occurs when brittle materials, such as cast iron, are cut and the chip breaks away from the work surface.

S

Saddle

(1) An H-shaped casting mounted on the top of the lathe ways and provides a means of mounting the cross-slide and the apron.

(2) The part of the horizontal milling machine that is fitted on top of the knee and may be moved in or out either manually or automatically.

Safety

The condition of being free from danger, injury, or damage.

Scraper

A hand tool used to move something hard, sharp, or rough across a surface, especially in order to clean it.

Screwdriver

A hand tool for fastening screws.

Scriber

A sharp instrument used to mark and lay out a pattern of work to be followed in subsequent machining operations.

Servo system

An automatic system for maintaining the read/write head on track. It can be either "open-loop", "quasi-closed-loop", or "closed-loop".

Set

The amount that the teeth are offset on either side of the center to produce clearance for the back of the band or blade.

Setup

Refers to line preparation to cut new width size for customer specification.

Shank

(1) The body of the tool bit or the part held in the tool holder.

(2) The part of the drill that fits into the holding device, whether it is a straight shank or a tapered shank.

Shaving

An operation in the manufacture of vitrified grinding wheels in which special wheel shapes and recesses are shaped or shaved to size in the green, or unburned, state on a shaving machine, which resembles a potter's wheel.

Shear angle

The angle of the area of the material where plastic deformation occurs.

Shear zone

The area where plastic deformation of the metal occurs.

Shell mill

A type of milling cutter that has cutting edges around its periphery and can be mounted on an arbor.

Side cutting edge angle

The angle the cutting edge forms with the side of the tool shank.

Side rake angle

The angle at which the face is ground away from the cutting edge.

Side relief angle

The angle ground on the flank of the tool below the cutting edge.

Slotting

A bandsaw operation where the saw makes a slit in the material.

Smelting

The process by which iron is removed from iron ore.

Spindle

(1) The hole through the headstock to which bar stock can be fed on a lathe.

(2) Part of the vertical milling machine in which cutting tools are installed.

(3) Part of the horizontal milling machine that provides the drive for arbors, cutters, and attachments used on the milling machine.

Spindle feed dial

The part of the horizontal milling machine that is set by a crank that is turned to regulate the spindle needle.

Split nut

Part of the lathe that when closed around the lead screw drives the carriage along by direct drive without using a clutch.

Splitting

A bandsaw operation where the saw divides the material into usually two pieces, especially lengthways.

Spot facing

The operation of smoothing and squaring the surface around a hole to provide a seat for the head of a cap screw or a nut.

Square head

Device used to check 45° and 90° angles and measure depths.

Squares

Instruments used to lay out lines at right angles to a machined edge to test the accuracy of surfaces that must be square, and to set up work for machining.

Standard

The level of quality or excellence that is accepted as the norm or by which actual attainments are judged.

Steady rest

Part of the lathe clamped to the ways that uses adjustable fingers to contact the workpiece and align it. It can be used in place of tailstock to support long or unstable parts being machined.

Steel rule

A scale used for measuring and layout.

Step pulleys

A part of the horizontal band saw used to vary the speed of the continuous blade to fit the type of material cut.

Stock

The material being machined. It can be any material and any shape. In the machine shop it usually refers to round or flat pieces of metal ready to be machined.

Surface gauge

A tool for scribing layout lines on a workpiece, or for transferring measurements from a rule to a workpiece.

Surface plate

Provides a precision reference surface for layout, checking, machining and gauging work.

Swiveling block

An attachment that enables the headstock to be tilted from 5° below the horizontal position to 10° beyond the vertical position.

Swivel table housing

The part of the horizontal milling machine that enables the table to be swiveled 45° to either side of the centerline.

T

Table

The part of the horizontal milling machine that supports the vise and the work.

Table handwheel

The part of the horizontal milling machine used to move the table toward or away from the column.

Tailstock

(1) Part of the lathe that fits on the inner ways of the bed and can slide towards any position on the headstock to fit the length of the work piece.

(2) Part of the universal dividing head set used in conjunction with the headstock to support work held between centers or the end of work held in a chuck.

Tailstock quill

Has a Morse taper to hold a lathe center, drill bit, or other tool.

Taper

A uniform change in the diameter of a workpiece measured along its axis.

Tapping

The operation of cutting internal threads in a hole with a cutting tool called a tap.

Taps

Cutting tools used to cut internal threads.

Tensile strength

The maximum amount of pull that a material will withstand before breaking.

Thread

A helical ridge of uniform section formed on the inside or outside of a cylinder or cone.

Three-dimensional shaping

A bandsaw operation where complicated shapes may be cut.

Tolerance

A range by which a product's gauge can deviate from those ordered and still meet the order's requirements.

Tool changer

A device that arranges multiple cutting tools in order and then positions these cutting tools for replacement in the machining center.

Tool function
A command that identifies a tool and calls for its selection.

Toolholder
The device used to rigidly hold a cutting tool in place. Toolholders are available in standardized sizes.

Tool length offset
The distance between the bottom of the fully retracted tool and the part Z0.

Tool offset
A correction entered for a tool's position parallel to a tool movement axis.

Tool post
Part of the lathe used to mount tool holders in which the cutting bits are clamped.

Toughness
The property of the metal to withstand shock or impact.

Trammel
Instrument used to scribe large arcs and circles.

Truing
An operation in the manufacture of vitrified grinding wheels in which the cured wheels are mounted in a special lathe and turned to the required size and shape by hardened-steel conical cutters, diamond tools, or special grinding wheels.

T-slot cutter
A cutting tool used to machine T-slot grooves in worktables, fixtures, and other holding devices.

Tumbling
A production process that is used for cleaning, polishing, and removing sharp corners and burrs from metal parts.

Turret
Part of the vertical milling machine that allows the milling head to be rotated around the column's center.

U

Universal chuck
A chuck having jaws which can be moved simultaneously so as to grasp objects of various sizes.

V

Vernier caliper
Precision tools used to make accurate measurements to within 0.001 inch or 0.02 mm.

Vertical bandsaw
A machine tool that provides a fast and economical method of cutting metal and other materials.

Vertical movement crank
Part of the vertical milling machine that moves the knee, saddle, and worktable up and down in unison.

Vise
A part of the horizontal band saw that can be adjusted to hold various sizes of workpieces.

W

Ways
The inner and outer guide rails on a lathe that are precision machined parallel to assure accuracy of movement.

Woodruff key
A semi-circular or half-round piece, resting in a circular groove cut in a shaft used for milling semi-cylindrical keyways in shafts.

Workholder
A device used to position and hold a workpiece in place.

Workpiece
A part that is being worked on. It may be subject to cutting, welding, forming, or other operations.

Worktable
The table that supports a workpiece during a manufacturing operation.

Worm wheel
A type of gear that engages with a worm to greatly reduce rotational speed or to allow higher torque to be transmitted.

Wrench
A hand tool with fixed or movable jaws, used to seize, turn, or twist objects such as nuts and bolts.

Appendix Ⅱ　MAHO 数控系统实验指导

Ⅱ.1　MAHO 数控系统

1. M 指令

M 指令见表Ⅱ.1。

表Ⅱ.1　M 指令

M 代码	功　能	M 代码	功　能
M00	程序停止		
M03	主轴顺时针旋转	M04	主轴逆时针旋转
M05	主轴停止，冷却液关	M06	自动换刀（铣）
M66	手动换刀（铣）	M67	假换刀（铣）
M08	冷却液开	M09	冷却液关
M18	主轴顺时针旋转，冷却液开	M14	主轴逆时针旋转，冷却液开
M30	程序结束并返回程序头		

2. 车削常用指令

Maho 车削常用 G 指令见表Ⅱ.2。系统支持 G 指令简写模式（如 G01 简写为 G1）。

表Ⅱ.2　Maho 车削常用 G 指令

G 指令	功　能	格　式
G00	快速移动	G00 X__ Z__
G01	直线移动	G01 X__（U__）Z__（W__）F__ S__ M__
G02/G03	顺/逆时针圆弧移动	G02/G03 X__ Z__ I__ K__（R__）F__ S__ M__
G04	暂停指令	G04 X__（单位 s）
G10	轴向容积切削加工	G10 X__（U__）Z__（W__）I__ K__ C__ F__
G11	径向容积切削加工	G11 X__（U__）Z__（W__）I__ K__ C__ F__
G12	轮廓精加工循环	G12 X__（U__）Z__（W__）F__
G13	容积切削加工循环调用	G13 N1=__ N2=__（执行）
G22	子程序调用	G22 N=__（En=__）

（续）

G 指令	功能	格式
G24	重复执行指令	G24 N1=__ N2=__ D__ (En=__)
G32	螺纹车削循环指令	G32 X__(U__) Z__(W__) C__ F__ (D__) (A__) (J__)(B__)F__
G33	单步螺纹车削循环指令	G33 X__(U__) Z__(W__) F__
G36/G37	车削/铣削功能指令	铣削动力头
G74	径向切槽定义循环	G74 X__(U__) Z__(W__) C__ A__ B__ Y__ R__ J__ P__ I__ K__ C__ F__
G75	轴向切槽调用循环	
G76	端面槽粗加工循环	G76 X__(U__)Z__(W__)C__ A__ B__ Y__ R__ J__ P__ I__ K__ C__ F__
G77	端面槽精加工循环	
G83	深孔钻削循环	G83 Z__(W__) C__ D__ J__ S1=__
G84	攻螺纹循环	G84 Z__(W__) I__ M1=__ S1=__
G92	零点偏移	G92 X__(U__) Z__(W__)　　（从基准点偏移）
G94	改变进给速度单位	G94 F__（单位 mm/min）
G95	改变进给速度单位	G94 F__（单位 mm/r）
G96	恒线速度切削	G96 S__ F__ D__
G97	取消恒线速度切削	G97 S__

3. T 指令

T 指令总是与 M6 指令或 M66、M67 配套使用。

格式：T__M6/ M66 / M67　　自动换刀/手动换刀/假换刀　（铣）
　　　T__　　　　　　　　　　　　　　　　　　　　　　（车）

4. S 指令

S 指令总是与 M3、M4、M5 指令或 M13、M14 配套使用。转速 20～6300r/min。

格式：S__M3/M4　　　　主轴以所给的转速正转/反转
　　　S__M13/M14　　　主轴以所给的转速正转/反转，冷却液开
　　　M5　　　　　　　主轴停止、冷却液关

5. F 指令

F 指令对除 G00 以外的任何移动或循环的指令起作用。范围 0～6000mm/min。

格式：F__

6. 铣削常用指令

Maho 铣削常用 G 指令见表 Ⅱ.3。

表Ⅱ.3 Maho 铣削常用 G 指令

G 指令	功能	格式
G00	快速移动	G00 X_ Y_ Z_ B_
G01	直线移动	G01 X_ Y_ Z_ B_ F_ S_ M_
G02/G03	顺/逆时针圆弧移动	G02/G03 X_ (Y_) Z_ I_ K_ (R_) F_ S_ M_
G14	重复执行指令	G14 N1=_ N2=_ J_
G17	XOY 工作平面定义	XOY(立式)，Z 轴为主轴方向
G18	XOZ 工作平面定义	XOZ(卧式)，Y 轴为主轴方向
G19	YOZ 工作平面定义	YOZ，X 轴为主轴方向
G40	取消刀具半径补偿	
G41	刀具半径左补偿	（沿刀具运动方向看，刀具在工件左边）
G42	刀具半径右补偿	（沿刀具运动方向看，刀具在工件右边）
G43	刀具靠近工件表面	G43 X_ (Z_) /G43 X_ (Y_)(G18/ G17 平面)
G44	刀具越过工件表面	G44 X_ (Z_) /G44 X_ (Y_)(G18/ G17 平面)
G51	机床坐标系零点	
G52	设置工件坐标系零点	
G53	取消工件坐标系	取消工件坐标系 G54～G59
G54～G59	设置工件坐标系零点	
G72	取消镜向/比例放缩	
G73	镜向/比例放缩指令	G73 X-1（Y-1）(Z-1)坐标轴号后跟-1，表示相应的坐标为反值 G73 A4=_ 放缩比例因子 A4
G79	循环调用	G79 X_ Y_ Z_ B_
G77	均布孔指令	G77 X_ Y_ Z_ R_ I_ J_ K_
G78	点位定义指令	G78 P_ X_ Y_ Z_
G81	钻削循环	G81 X_ Y_ Z_ B_
G83	深孔钻削循环	G83 X_ Y_ Z_ B_ I_ J_ K_
G84	攻丝循环	G84 X_ Y_ Z_ B_ I_ J_ S_ （或 F_）
G85	铰削循环	G85 X_ Y_ Z_ B_
G86	镗削循环	G86 X_ Y_ Z_ B_
G87	方腔铣削循环	G87 X_ Y_ Z_ B_ R_ I_ J_ K_
G88	键槽铣削循环	G88 X_ Y_ Z_ B_ J_ K_
G89	圆腔铣削循环	G89 Z_ B_ R_ I_ J_ K_
G90	绝对坐标	

（续）

G 指令	功能	格式
G91	增量坐标	（部分文献译为：相对坐标）
G92	增量坐标系变换指令	G92 X_ Y_ Z_ B1=_ L1=_ B4=_
G93	绝对坐标系变换指令	G93 X_ Y_ Z_ B2=_ L2=_ B4=_
G98	图形窗口定义指令	G98 X_ Y_ Z_ I_ J_ K_ （XYZ 起始坐标，IJK 平行于 XYZ）
G99	图形毛坯定义指令	G99 X_ Y_ Z_ I_ J_ K_ （车削 X_ Z_ I_ 外径/长度/内径）

Ⅱ.2　MAHO GR350C 车削编程

1. G00 快速移动（定位）指令

格式：G00　X__Z__

说明：刀具以快速移动速度从当前点运动到坐标指定的位置（快速移动速度为 6m/min）。

G00 指令执行时，主轴可以不转动。

G00 快速移动指令如图Ⅱ.1 所示。

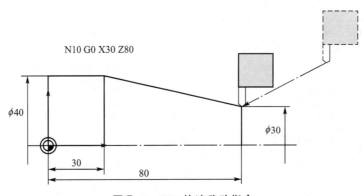

图Ⅱ.1　G00 快速移动指令

2. G01 直线移动（线性插补）指令

格式：G01　X__(U__)Z__(W__)F__S__M__

刀具以给定的进给速度、转速，从当前点移动到坐标所指定的点。运动时两个坐标同时移动，同时到达终点。

正常车削时，主轴采用 M4 指令（正好与铣床相反），从尾架向主轴方向看，主轴按顺时针旋转。

U、W 分别代表 X 和 Z 方向上的相对位移。

G01 直线移动指令如图Ⅱ.2 所示。

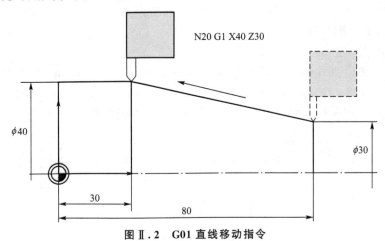

图Ⅱ.2　G01 直线移动指令

车床在编程时可采用半径编程或直径编程，通常选用直径编程。如无特殊说明，均采用直径编程。

3. G02/G03 顺时针/逆时针圆弧移动（圆弧插补）指令

格式：G02(G03) X__Z__I__K__(R__)F__S__M__

说明：刀具从当前点顺/逆时针运动，以给定的圆弧中心坐标（I、K）或圆弧半径 R，移动到坐标给定的目标点。

圆弧插补指令如图Ⅱ.3 所示。

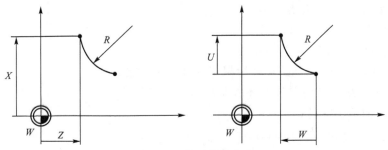

图Ⅱ.3　圆弧插补指令

4. G04 暂停指令

格式：G04 X__

说明：暂停给定的时间（s），常用于切槽时，底部停留。

5. G10 轴向容积切削加工循环

格式：G10 X__(U__) Z__(W__) I__ K__ C__ (F__)

X——起始点 S 的 X 绝对坐标；

U——起始点 S 的 X 增量坐标；

Z——起始点 S 的 Z 绝对坐标；

W——起始点 S 的 Z 增量坐标；

I——X 方向为 G12 精加工循环所保留的精加工余量；

K——Z 方向为 G12 精加工循环所保留的精加工余量；

C——每次的切削深度；

F——进给量，通常为 mm/r。

轮廓定义必须以直线运动移动到轮廓的第一点 A（图Ⅱ.4）。轮廓的描述可用基本 G 指令，如 G01、G02、G03，或几何描述指令 G15、G16、G17、G18。编程坐标可用增量坐标和绝对坐标。如果需要恒速切削模式，可在 G10 语句之前加上 G96。

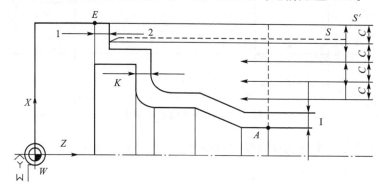

图Ⅱ.4　G10 轴向容积切削加工循环指令

6. G11 径向容积切削加工循环

格式：G11 X＿(U＿)　Z＿(W＿)　I＿　K＿　C＿　(F＿)

X——起始点 S 的 X 绝对坐标；

U——起始点 S 的 X 增量坐标；

Z——起始点 S 的 Z 绝对坐标；

W——起始点 S 的 Z 增量坐标；

I——X 方向为 G12 精加工循环所保留的精加工余量；

K——Z 方向为 G12 精加工循环所保留的精加工余量；

C——每次的切削深度；

F——进给量，通常为 mm/r。

G11 径向容积切削加工循环指令如图Ⅱ.5 所示。

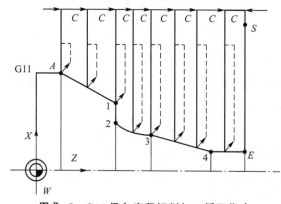

图Ⅱ.5　G11 径向容积切削加工循环指令

7. G12 轮廓精加工循环

格式：G12 X__(U__) Z__(W__) (F__)
X——起始点 S 的 X 绝对坐标；
U——起始点 S 的 X 增量坐标；
Z——起始点 S 的 Z 绝对坐标；
W——起始点 S 的 Z 增量坐标。

8. G13 容积切削加工循环调用（执行）

格式：G13 N1=__ N2=__
N1——轮廓定义的第一条语句；
N2——轮廓定义的最后一条语句。
说明：轮廓的加工与调用。

9. G96/G97 选择/取消恒线速度切削指令

格式：G96 S__ F__ D__
S——切削线速度(m/min)；
F——进给速度；
D——主轴最高限速(r/min)。
G97 S__
S——主轴转数(r/min)
说明：恒速切削时，保持线速度不变，直径越小，转数越高，但最高不超过最高限速。

G10 加工循环实例（图Ⅱ.6）：

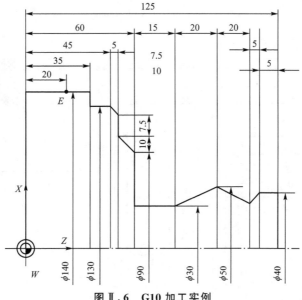

图Ⅱ.6　G10 加工实例

% PM9000001
N9000001
N1 G54;　　　　　　　　　　　　　　　（工作坐标系）

N2 G99 X140 Z127;

N3 G96 S100 D2500 T1013 M4;　　　　　　（恒线速度切削模式，线速度100m/min，最高转速不超过2500r/min）

N4 G10 X145 Z130 I0.5 K0.5 C2.5 F0.5;　　（循环定义，精加工余量X、Z方向均为0.5mm，每次切削厚度2.5mm，轮廓起点(145，130)）

N8001 G01 X40 Z125;　　　　　　　　　　（轮廓定义开始）

N5 G01 W-5;

N6 G01 X30 W-5;

N7 G01 X50 Z95;

N8 G01 X30 Z75;

N9 G01 Z60;

N10 G01 X90;

N11 G01 U20 Z50;

N12 G01 U15;

N13 G01 X130 Z45;

N14 G01 Z35;

N15 G01 X140;

N16 G01 Z28;　　　　　　　　　　　　　（轮廓定义结束）

N17 G13 N1= 8001 N2= 16;　　　　　　　（调用容积切削循环）

N18 G00 X200 Z145;　　　　　　　　　　（退刀）

N19 T2023;　　　　　　　　　　　　　　（换刀）

N20 G12 X145 Z127 S200;　　　　　　　　（精加工循环定义）

N21 G13 N1= 8001 N2= 16;　　　　　　　（调用容积切削循环）

N22 G00 X300 Z350;　　　　　　　　　　（退刀）

N23 M30;　　　　　　　　　　　　　　　（程序结束）

G11加工循环实例（图Ⅱ.7）：

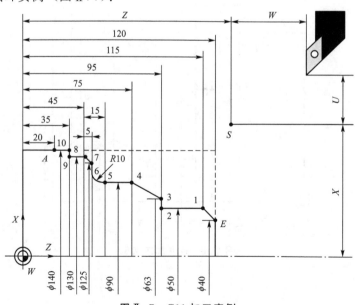

图Ⅱ.7　G11加工实例

```
% PM9000002
N9000002
N1 G54;                              （工作坐标系）
N2 G99 X140 Z127;
N3 G96 S100 D2500 T1013 M4;          （恒速切削模式，线速度100m/min，最高转速不超过
                                      2500r/min）
N4 G00 X200 Z145;                    （快速移动到起始点）
N5 G11 X145 Z120 I0.5 K0.5 C2.5 F0.5;
                                     （循环定义，精加工余量X、Z方向均为0.5mm，每次切
                                      削厚度2.5mm，轮廓起点(145mm，120mm)）
N6 G01 Z25;                          （轮廓定义开始）
N7 G41 X140;
N8 G01 Z35;
N9 G01 X130;
N10 G01 Z45;
N11 G01 X125 Z50;
N12 G01 X110;
N13 G03 X90 Z60 R10;
N14 G01 Z75;
N15 G01 X63 Z95 F0.1;
N16 G01 X50;
N17 G01 Z115;
N18 G01 X40 Z120;
N19 G40;                             （轮廓定义结束）
N20 G13 N1= 6 N2= 19;                （调用容积切削循环）
N21 G00 X200 Z145;                   （退刀）
N22 T2023;                           （换刀）
N23 G12 X145 Z120 S200;              （精加工循环定义）
N24 G13 N1= 6 N2= 19;                （调用容积切削循环）
N25 G00 X300 Z350;
N26 M30;                             （退刀，程序结束）
```

10. G22 子程序调用指令

格式：G22 N= __ （En= __ ）

N——子程序号；

En——子程序调用条件，当该数值大于或等于0时，执行子程序，n从0到255。

例：

N150   G22   E60   N＝9100   （当E60≥0时，控制转换到9100）

11. G24 重复执行指令

格式：G24 N1= __ N2= __ D__ （En= __ ）

N1——重复执行语句的起始号；

N2——重复执行语句的终止号；

D——重复次数；

En——子程序调用条件，当该数值大于或等于0时，执行子程序，n从0到255。

12．G32 螺纹车削循环指令

格式：G32 X＿(U＿) Z＿(W＿) C＿(D＿) (A＿) (J＿)(B＿) F＿

X——螺纹底径；

Z——坐标终点，相对编程零点；

U——螺纹深度(从起始点算起)，＋U内螺纹，－U外螺纹；

W——螺纹长度(从起始点算起)，可用±号；

C——第一次切削深度，如果C＝U，只切削一次；如果C＜U，则切削多次；

D——螺纹精加工深度；

A——刀尖角的一半；

J——螺纹切出锥度的终止直径；

B——锥度比例；

F——螺纹螺距。

G32螺纹车削循环指令如图Ⅱ.8所示。

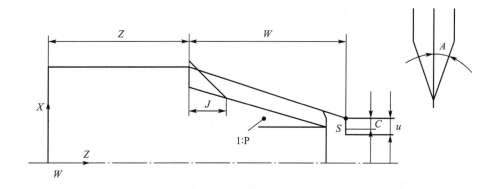

图Ⅱ.8　G32 螺纹车削循环指令

如果J＝0或缺省，螺纹将仅切削到终点处，无切出锥度。如果J≠0，螺纹切至终点前，再以锥度切出，直到终点的Z坐标。锥度比例B为锥度1：P。

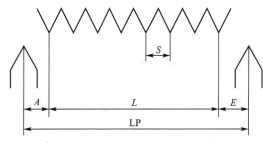

图Ⅱ.9　G33 单步螺纹车削循环指令

13．G33 单步螺纹车削循环指令

格式：G33 X＿(U＿) Z＿(W＿) F＿

说明：用于加工普通螺纹、锥螺纹、多头螺纹和多刀切削螺纹。

螺纹总长LP应包括螺纹导入长度E、螺纹长度L和螺纹导出长度A。

G33单步螺纹车削循环指令如图Ⅱ.9所示。

螺纹车削实例（图Ⅱ.10）：

N6 G00 X64 Z155;
N7 G33 X61.835 Z40 F2;

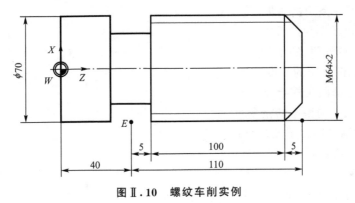

图Ⅱ.10 螺纹车削实例

锥螺纹车削实例（图Ⅱ.11），螺距为3。

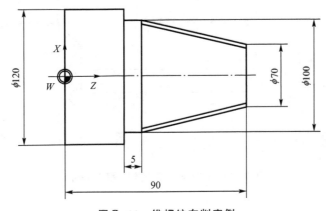

图Ⅱ.11 锥螺纹车削实例

N6 G00 X66.752 Z95;
N7 G33 X96.752 Z45 F3;
N8 G00 X130;

14．G41/G42 刀具半径左(右)补偿功能指令（G40 刀具半径补偿功能取消指令）

格式：G41/G42

15．G94/G95 改变进给速度单位指令

格式：G94/G95 F__（G94 常用进给速度单位为 mm/min；G95 常用进给速度单位为 mm/r）

16．G74 径向切槽定义循环指令

格式：G74 X__(U__) Z__(W__) A_ B_ Y_ R_ J_ P_ I_ K_ C_ F__
X——槽底部直径(或半径)；
U——槽深；
Z——槽终点坐标；

W——槽宽；

A——槽第一条斜边的角度(°)；

B——槽最后一条边的角度；

Y——在槽顶部倒角的长度；

R——在槽顶部圆角的半径；

J——在槽底部倒角的长度；

P——在槽底部圆角的半径；

I——沿 X 轴方向留下的余量；

K——沿 Z 轴方向留下的余量；

C——刀具宽度；

F——进给速度。

G74 径向切槽加工循环指令如图Ⅱ.12 所示。

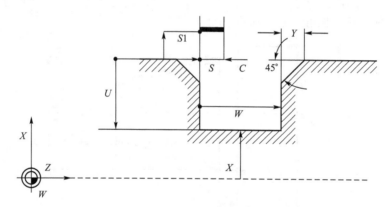

图Ⅱ.12　G74 径向切槽加工循环指令

切槽实例（图Ⅱ.13）：

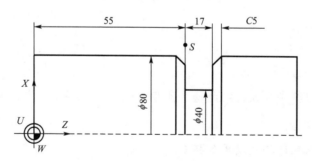

图Ⅱ.13　切槽实例

N10 G01 X80 Z56;
N20 G74 X40 W17 Y5 C6 I1 K1;

17. G76 端面槽粗加工循环指令

格式：G76 X_(U_) Z_(W_) A_ B_ Y_ R_ J_ P_ I_ K_ C_ F_

X——槽底部直径(或半径)；

U——槽深；
Z——槽终点坐标；
W——槽宽；
A——槽第一条斜边的角度(°)；
B——槽最后一条边的角度；
Y——在槽顶部倒角的长度；
R——在槽顶部圆角的半径；
J——在槽底部倒角的长度；
P——在槽底部圆角的半径；
I——沿 X 轴方向留下的余量；
K——沿 Z 轴方向留下的余量；
C——刀具宽度；
F——进给速度。

18. G92 零点偏移指令

格式：G92 X__ Z__   （从基准点偏移）
　或 G92 U__ W__   （从另一个基准点偏移）

G92 零点偏移实例如图Ⅱ.14 所示。

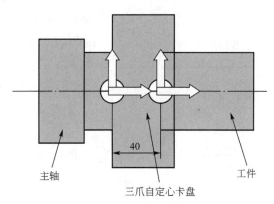

图Ⅱ.14 G92 零点偏移实例

N50 G92 Z40;

19. G51/G52 零点偏移取消/建立指令

格式：G51/G52

N75 G51;   取消零点偏移
N76 G92 X0 Z192;   建立新基准点位置

格式：G53/G54~ G59
说明：G53/G54～G59 零点偏移取消/激活指令。

20. G70/G71 英制/公制指令

格式：G70   英制（单位为 in）

G71　　　公制(单位为 mm)

21. G36/G37 车削/铣削功能指令

用于铣削和车削功能切换。

Ⅱ.3　MAHO 600C 铣削编程

1. G00 快速移动(定位)指令

格式：G00　X__ Y__ Z__ B__

说明：刀具以快速移动速度从当前点运动到坐标指定的位置(快速移动速度为 6m/min)。当移动涉及 3 个坐标时，按下列顺序移动坐标轴：

在 *XOZ* 平面内加工时(G18，即立式主轴)，刀具先移动 *X*、*Z* 轴，再移动 *Y* 轴；在 *XOY* 平面内加工时(G17，即卧式主轴)，刀具先移动 *X*、*Y* 轴，再移动 *Z* 轴。

2. G01 直线移动(线性插补)指令

格式：G01　X__ Y__ Z__ B__ F__ S__ M__

说明：刀具以给定的进给速度、转速，从当前点移动到坐标所指定的点。运动时 4 个坐标同时移动，同时到达终点，G01 可以简写为 G1。G01 直线移动指令如图Ⅱ.15 所示。

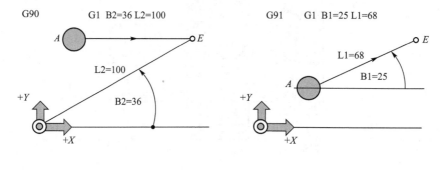

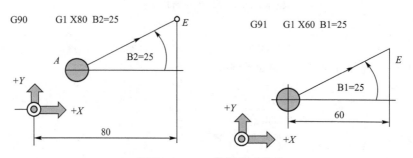

图Ⅱ.15　G01 直线移动指令

G01 指令的目标点坐标也可用极坐标来表示。在绝对坐标下，用 L2、B2 代表长度和与水平轴的夹角；在增量坐标下，用 L1、B1 代表长度和与水平轴的夹角。也可用 *X* 坐标和 B1 表示。

3. G02、G03 顺时针、逆时针圆弧移动(圆弧插补)指令

格式：G02(G03) X__(Y__)Z__ I__K__(R__)F__S__M__

说明：刀具从当前点顺(逆时针)运动，以给定的圆弧中心坐标(I、K)或圆弧半径R，移动到坐标给定的目标点。

刀具坐标使用X—Y组合、X—Z组合，还是Y—Z组合，取决于所选定的工作平面(G17、G18、G19)。圆心坐标也相应为I—J、I—K或J—K。

圆弧夹角大于180°与小于180°时，使用的方法有所不同。

圆弧移动指令如图Ⅱ.16所示。

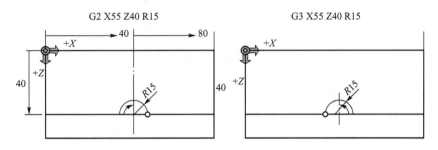

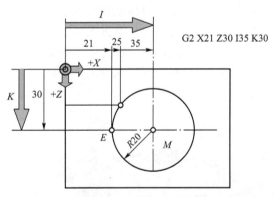

图Ⅱ.16 圆弧移动指令

整圆的起点与终点相同。G02/G03命令还可用来加工螺旋线。

4. G17、G18、G19 工作平面定义指令

格式：G17(G18、G19) 无参数

G17定义工作平面为XOY(立式)，Z轴为主轴方向；
G18定义工作平面为XOZ(卧式)，Y轴为主轴方向；
G19定义工作平面为YOZ，X轴为主轴方向。

5. G90、G91 绝对坐标、增量坐标指令

格式：G90(G91) 无参数

G90：出现该指令后所有的坐标值均为绝对值坐标(相对于工作坐标系)，直到遇到G91指令取消，机床开机后缺省状态为G90。

G91：出现该指令后所有的坐标值均为增量值坐标（相对于前一点），直到遇到 G90 指令取消。

绝对坐标、增量坐标指令如图Ⅱ.17 所示。

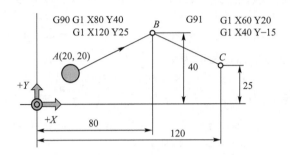

图Ⅱ.17　绝对坐标、增量坐标指令

6. G40、G41、G42、G43、G44 刀具半径补偿指令

格式：G40　　取消刀具半径补偿；
　　　G41　　刀具半径左补偿（沿刀具运动方向看，刀具在工件左边）；
　　　G42　　刀具半径右补偿（沿刀具运动方向看，刀具在工件右边）；
　　　G43　X_（Z_）　　G18 平面：刀具靠近工件表面；
　　　G43　X_（Y_）　　G17 平面：刀具靠近工件表面；
　　　G44　X_（Z_）　　G18 平面：刀具越过工件表面；
　　　G44　X_（Y_）　　G17 平面：刀具越过工件表面。

刀具半径补偿指令如图Ⅱ.18 所示。

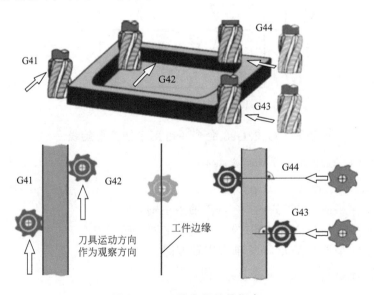

图Ⅱ.18　刀具半径补偿指令

例：铣通槽（图Ⅱ.19），板厚 20mm。

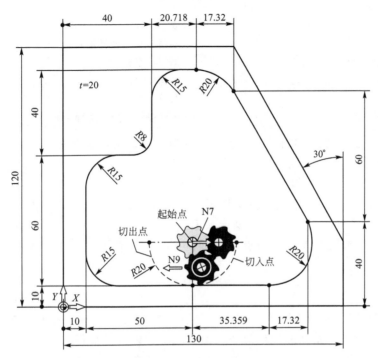

图Ⅱ.19 铣通槽实例图

```
% PM9001
N9001                              （铣通槽实例）
N1 G17 S900 T31 M66;               （铣刀直径 φ10mm）
N2 G54;                            （工件坐标零点）
N3 G98 X-10 Y-10 Z-20 I150 J140 K30;  （模拟窗口）
N4 G99 X0 Y0 Z-20 I130 J120 K20;   （毛坯大小）
N5 G00 X60 Y30 Z8 M3;              （快速运动到起始点）
N6 G01 Z-21 F50;                   （进刀到给定深度）
N7 G43 X80 F100;                   （刀具半径补偿，靠近）
N8 G42;                            （刀具半径右补偿）
N9 G02 X60 Y10 R20;                （或 I60 J30）
N10 G01 X25;
N11 G02 X10 Y25 R15;               （或 I25 J25）
N12 G01 Y55;
N13 G02 X25 Y70 R15;               （或 I25 J55）
N14 G01 X32;
N15 G03 X40 Y78 R8;                （或 I32 J78）
N16 G01 Y95;
N17 G02 X55 Y110 R15;              （或 I55 J95）
N18 G01 X60.718;
N19 G02 X78.039 Y100 R20;          （或 I60.718 J90）
N20 G01 X112.679 Y40;
N21 G02 X95.359 Y10 R20;           （或 I90.359 J30）
```

N22 G01 X60;
N23 G02 X40 Y30 R20;   （或 I60 J30）
N24 G40;   （取消刀具半径补偿）
N25 G00 Z50 M5;   （退刀，主轴停）
N26 G53;   （取消 G54）
N27 T0 M66;   （取下刀具）
N28 M30;   （程序停止，返回程序头）

7. G51、G52、G53、G54—G59 机床零点、工件零点、参考点

格式：G51   机床坐标系零点
　　　G52   设置工件坐标系零点
　　　G53   取消工件坐标系 G54～G59
　　　G54～G59   设置工件坐标系零点

机床开机后，首先要执行参考点，否则机床与操作有关的按钮与菜单均无效（为安全起见）。

8. 循环指令

对于经常使用的一系列加工动作，如钻孔、钻深孔、镗孔、铰孔、铣削圆腔、方腔等，数控系统提供了一系列的循环指令（图Ⅱ.20），以方便编程。

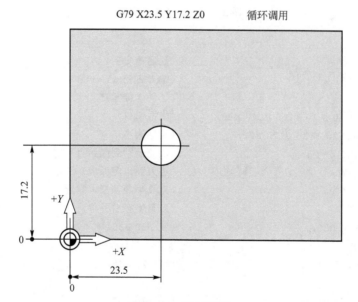

图Ⅱ.20　循环调用

循环指令分成两类：循环定义指令和循环调用指令。

前者定义了加工循环所必需的一些参数，如 G81 钻孔循环、G84 攻螺纹循环等；后者则表示在何处执行该循环，如 G79、G77 均布孔等。

1) G81 钻削循环指令

格式：G81 X__ Y__ Z__ B__

X——钻孔到底部时停留的时间(s)；

Y——安全距离，加工时刀具快速移动到离工件表面 Y mm 处，再开始工进；

Z——孔深，为负数；

B——退刀距离。

G81 钻孔循环指令如图Ⅱ.21 所示。

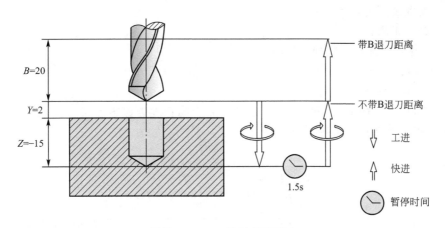

图Ⅱ.21　G81 钻孔循环指令

当工件表面较平整时，且不会与表面上的夹具发生干涉时，可用较小的安全距离；反之要选用较大的安全距离。

2) G83 深孔钻削循环指令

格式：G83 X__ Y__ Z__ B__ I__ J__ K__

X——钻孔到底部时停留的时间(s)；

Y——安全距离，加工时刀具快速移动到离工件表面 Y mm 处，再开始工进；

Z——孔深，为负数；

B——退刀距离；

I——深度递减量；

J——每次退刀量，若无此项，则退回至安全平面；

K——第一次进刀深度。

相比之下，退刀至安全平面断屑效果较好，用于较深的孔。

G83 深孔钻削循环指令如图Ⅱ.22 所示。

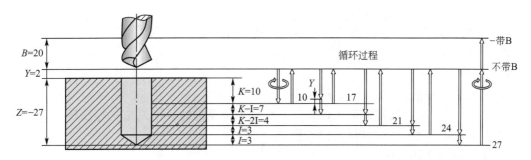

图Ⅱ.22　G83 深孔钻削循环指令

3) G84 攻螺纹循环指令

格式：G84 X__ Y__ Z__ B__ I__ J__ S__（或 F__）

X——攻螺纹到底部时停留的时间(s)；

Y——安全距离，加工时刀具快速移动到离工件表面 Y mm 处，再开始工进；

Z——攻螺纹深度，为负数；

B——退刀距离；

I——在螺纹底部逐步降速的圈数；

J——螺距；

S——主轴转速；

F——进给量，转速 S、进给量 F 和螺距 J 之间存在着以下关系：$F = S \times J$。

G84 攻螺纹循环指令如图Ⅱ.23所示。

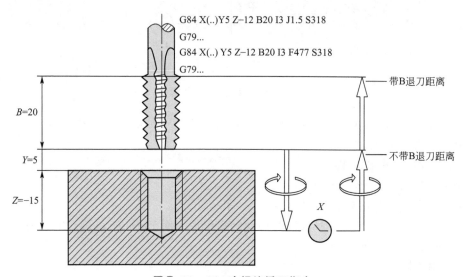

图Ⅱ.23　G84 攻螺纹循环指令

攻螺纹循环一旦开始，按通常的停止按钮和进给量与主轴转速调节旋钮，并不起作用，而是要等循环结束后才停止。因此，在加工中要特别注意。

4) G85 铰削循环指令

格式：G85 X__ Y__ Z__ B__

X——铰孔到底部时停留的时间(s)；

Y——安全距离，加工时刀具快速移动到离工件表面 Y mm 处，再开始工进；

Z——铰孔深度，为负数；

B——退刀距离。

G85 铰削循环指令如图Ⅱ.24所示。

铰孔时，进入铰孔循环后为工进深度，铰孔完毕退出时，不像钻孔等循环一样快速退回，而是以工进速度退回，以保证铰孔质量。

5) G86 镗削循环指令

格式：G86 X__ Y__ Z__ B__

X——镗孔到底部时停留的时间(s)；

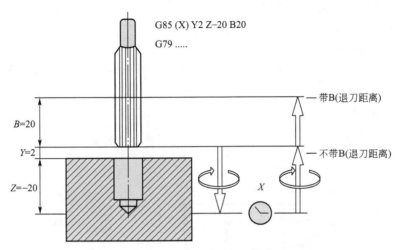

图Ⅱ.24　G85 铰削循环指令

Y——安全距离，加工时刀具快速移动到离工件表面 Ymm 处，再开始工进；
Z——镗孔深度，为负数；
B——退刀距离。

镗孔完毕退出时，主轴停止转动，快速退回。
G86 镗削循环指令如图Ⅱ.25 所示。

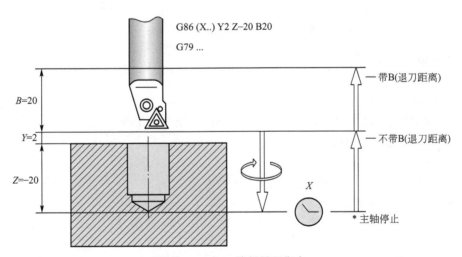

图Ⅱ.25　G86 镗削循环指令

6）G87 方腔铣削循环指令
格式：G87 X__Y__Z__B__R__I__J__K__
X——方腔沿 X 方向的长度；
Y——方腔沿 Y（或 Z）方向的长度；
Z——铣削深度，为负数；
B——退刀距离；
R——拐角半径（必须大于刀具半径）；

I——在刀具宽度方向上，刀具切入量与刀径之比，缺省为83％；
J——加工方向，J1 为顺铣，J-1 为逆铣；
K——每次铣削深度。

沿深度方向进给时，进给速度为 F 的一半。用 G79 调用时，坐标值为方腔上表面中心。

G87 方腔铣削循环指令如图Ⅱ.26 所示。

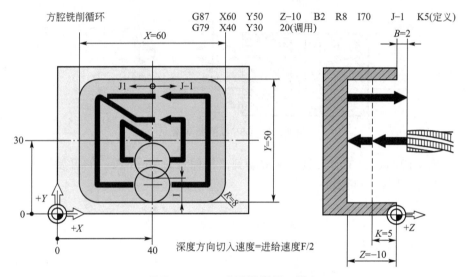

图Ⅱ.26　G87 方腔铣削循环指令

7）G88 键槽铣削循环指令

格式：G88 X__ Y__ Z__ B__ J__ K__

X——键槽沿 X 方向的长度；

Y——键槽沿 Y（或 Z）方向的长度，刀具半径必须小于 Y/2；

Z——铣削深度，为负数；

B——退刀距离；

J——加工方向，J1 为顺铣，J-1 为逆铣；

K——每次铣削深度。

沿深度方向进给时，进给速度为 F 的一半。用 G79 调用时，坐标值为键槽上表面左侧圆弧中心。

G88 键槽铣削循环指令如图Ⅱ.27 所示。

8）G89 圆腔铣削循环指令

格式：G89 Z__ B__ R__ I__ J__ K__

Z——铣削深度，为负数；

B——退刀距离；

R——圆腔半径，刀具半径必须小于圆腔半径；

I——在刀具宽度方向上，刀具切入量与刀径之比，缺省为83％；

J——加工方向，J1 为顺铣，J-1 为逆铣；

K——每次铣削深度。

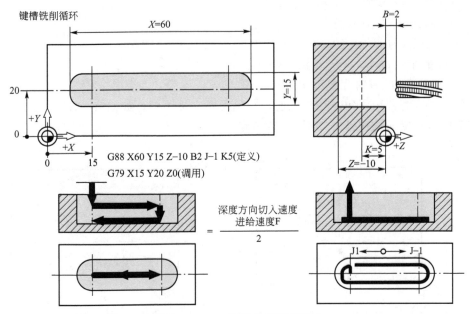

图Ⅱ.27 G88 键槽铣削循环指令

G89 圆腔铣削循环指令如图Ⅱ.28 所示。

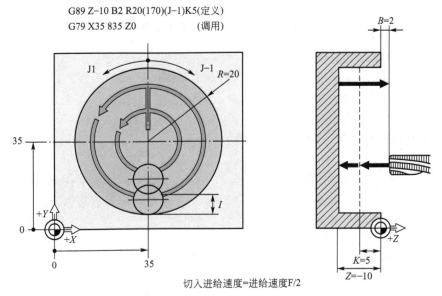

图Ⅱ.28 G89 圆腔铣削循环指令

9) G78 点位定义指令

格式：G78 P__ X__ Y__ Z__

说明：用 P__ 代表 X__ Y__ Z__。

10) G77 均布孔定义调用指令

格式：G77 X__ Y__ Z__ R__ I__ J__ K__

X__Y__Z__——均布孔的中心点坐标;

R——均布孔分布圆半径;

I——起始角(与 X 轴正方向);

J——孔的个数;

K——终止角(缺省时为整圆均布)。

说明:G77 为调用语句,在其之前,它仍需要定义语句,如 G81~G89。

G77 均布孔定义调用指令如图Ⅱ.29 所示。

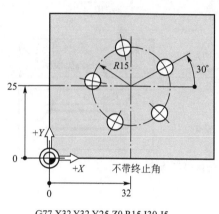
G77 X32 Y32 Y25 Z0 R15 I30 J5

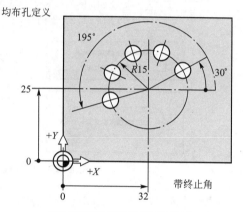
G77 X32 Y25 Z0 R15 I30 J5 K195

图Ⅱ.29 G77 均布孔定义调用指令

11) G14 程序段重复指令

格式:G14 N1=__ N2=__ J__

N1——重复执行的起始程序号;

N2——重复执行的终止程序号;

J——重复执行的次数(缺省为重复 1 次)。

12) G92、G93 坐标系变换指令

格式:G92 X__Y__Z__B1=__L1=__B4=__   增量坐标零点平移/旋转

G93 X__Y__Z__B2=__L2=__B4=__   绝对坐标零点平移/旋转

X、Y、Z——直线坐标的平移量;

B1、L1——极坐标(G92);

B2、L2——极坐标(G93);

B4——旋转后坐标轴与旋转前的坐标轴的夹角。

坐标系变换指令如图Ⅱ.30 所示。

有时,为了编程方便,可对工件坐标系零点进行变换,使编程在新的坐标系内进行。坐标系变换可分为绝对坐标系变换 G93 和增量坐标系变换 G92 两种。

坐标系变换实例(图Ⅱ.31),铣槽深 20mm,板厚 20mm。

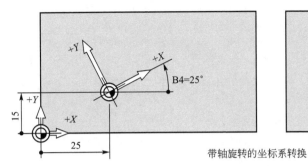
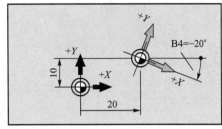

带轴旋转的坐标系转换

图Ⅱ.30 坐标系变换指令

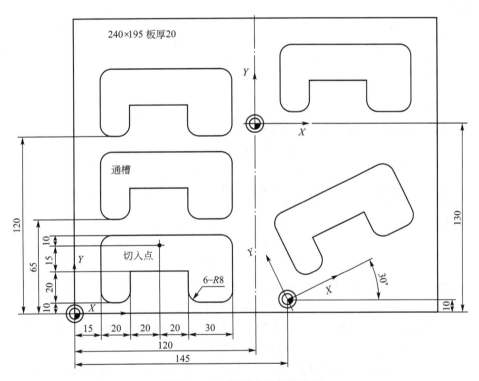

图Ⅱ.31 坐标系变换编程实例

```
% PM9002
N9002                        （坐标系变换的实例）
N1 G17 S800 T31 M66;         （铣刀直径 φ10mm）
N2 G54;
N3 G98 X-10 Y-10 Z-10 I260 J215 K30;
N4 G99 X0 Y0 Z-20 I240 J195 K20;
N5 G00 X55 Y45 Z2 M13;
N6 G01 Z-21 F50;             （进刀）
N7 G43 Y55 F100;             （靠近 Y55）
```

N8 G42;   （刀具半径右补偿）
N9 G01 X97;   （加工轮廓）
N10 G02 X105 Y47 R8;
N11 G01 Y18;
N12 G02 X97 Y10 R8;
N13 G01 X83;
N14 G02 X75 Y18 R8;
N15 G01 Y30;
N16 G01 X35;
N17 G01 Y18;
N18 G02 X27 Y10 R8;
N19 G01 X23;
N20 G02 X15 Y18 R8;
N21 G01 Y47;
N22 G02 X23 Y55 R8;
N23 G01 X55;   （轮廓加工结束）
N24 G00 Z50;
N25 G40;   （取消刀具半径补偿）
N26 G92 Y55;   （增量坐标系变换）
N27 G14 N1= 5 N2= 26 J2;   （重复循环 2 次）
N28 G93 X120 Y130;   （绝对坐标系变换）
N29 G14 N1= 5 N2= 25;   （重复循环 1 次）
N30 G93 X145 Y10 B4= 30;   （绝对坐标系变换）
N31 G14 N1= 5 N2= 25;   （重复循环 1 次）
N32 G00 Z50 M5;   （退刀）
N33 G53;   （取消 G54）
N34 M30;   （程序结束，并回绕）

13) G72、G73 镜向/比例放缩指令

格式：G72   无参数，取消 G73；
     G73 X-1（Y-1）(Z-1)   坐标轴号后跟-1，表示相应的坐标为相反值；
     G73 A4= ___   放缩比例因子。

说明：在实际应用中，经常会遇到形状相同、但旋转了一定角度的零件，或按一定比例进行放大缩小的零件。这时，就要使用到 G72、G73 指令。

镜向指令如图Ⅱ.32 所示。

G72、G73 指令仅仅是一条说明指令，说明了从此时起，坐标轴的对称关系或比例关系，它本身并不是可执行指令。所以，还要结合 G14 重复指令等来完成对称零件或相似零件的加工。

镜向实例（图Ⅱ.33）：

简单工艺流程：毛坯 240mm×130mm×5mm，45 钢，工件坐标系原点如上图。

(1) 下料 240mm×130mm×5mm；

(2) 铣右上第一个槽，切入点(55, 45)；铣刀 ϕ10mm；

(3) Y 轴镜像，加工左上槽；

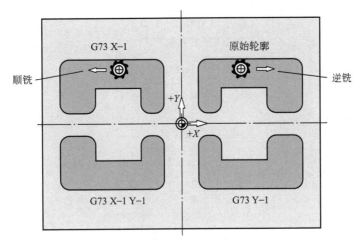

图Ⅱ.32 镜向指令

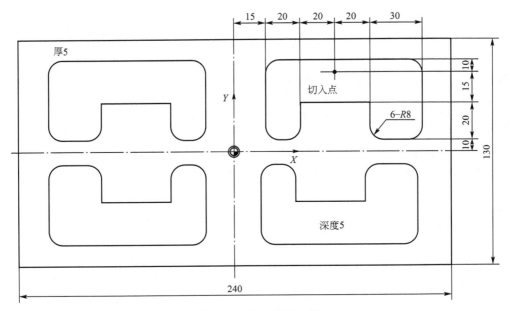

图Ⅱ.33 镜向编程实例

(4) 原点镜像,加工左下槽;
(5) X轴镜像,加工右下槽;
(6) 退刀。

% PM9003
N9003　　　　　　　　　　　(镜向的实例,毛坯 240mm×130mm×5mm)
N1 G17 S800 T31 M66;　　(铣刀直径 φ10mm)
N2 G54;
N3 G98 X-130 Y-75 Z-10 I260 J150 K20;
N4 G99 X-120 Y-65 Z-5 I240 J130 K5;
N5 G0 X55 Y45 Z2 M13;　　(G00 简写为 G0)

N6 G1 Z-6 F50;   （进刀）
N7 G43 Y55 F100;   （靠近 Y55）
N8 G42;   （刀具半径右补偿）
N9 G1 X97;   （加工轮廓）
N10 G2 X105 Y47 R8;
N11 G1 Y18;
N12 G2 X97 Y10 R8;
N13 G1 X83;
N14 G2 X75 Y18 R8;
N15 G1 Y30;
N16 G1 X35;
N17 G1 Y18;
N18 G2 X27 Y10 R8;
N19 G1 X23;
N20 G2 X15 Y18 R8;
N21 G1 Y47;
N22 G2 X23 Y55 R8;
N23 G1 X55;   （轮廓加工结束）
N24 G0 Z50;
N25 G40;   （取消刀具半径补偿）
N26 G73 X-1;   （对 Y 轴作镜向）
N27 G14 N1= 5 N2= 25;   （重复 1 次）
N28 G72;
N29 G73 X-1 Y-1;   （对原点作中心对称）
N30 G14 N1= 5 N2= 25;   （重复 1 次）
N31 G72;
N32 G73 Y-1;   （对 X 轴作镜向）
N33 G14 N1= 5 N2= 25;   （重复 1 次）
N34 G72;   （取消 G73）
N35 G0 Z50 M5;
N36 G53;
N37 M30;

Appendix Ⅲ　FANUC 数控系统实验指导

Ⅲ.1　FANUC 数控系统

FANUC 数控系统编程规则

(1) 小数点编程：在输入的任何坐标字(包括 X、Y、Z、I、J、K、U、W、R 等)数值后须加小数点，即 X100 须记作 X100.0。否则系统认为坐标字数值为脉冲值，100×0.001mm＝0.1mm。

(2) 绝对坐标值方式与增量值坐标方式：用 U 或 W 表示增量坐标。在程序段出现 U 即表示 X 方向的增量值，出现 W 即表示 Z 方向的增量值。同时，允许绝对坐标值方式与增量值坐标方式混合编程。

(3) 进给功能：系统默认进给方式为每转进给(毫米/转)。

(4) 程序名的指定：本系统程序名采用字母 O 后跟四位数字的格式。子程序文件名遵循同样的命名规则。通常在程序开始指定文件名。程序结束须加 M30 或 M02 指令。

(5) G 指令简写模式：系统支持 G 指令简写模式(如 G01 简写为 G1)。

常用文字码及其含义见表Ⅲ.1，M 代码见表Ⅲ.2。

表Ⅲ.1　常用文字码及其含义

功能	文字码	含义
程序号	O：ISO/：EIA	表示程序名代号(1~9999)
程序段号	N	表示程序段代号(1~9999)
准备机能	G	确定移动方式等准备功能
坐标字	X、Y、Z、A、C	坐标轴移动指令(±99999.999mm)
坐标字	R	圆弧半径(±99999.999mm)
坐标字	I、J、K	圆弧圆心坐标(±99999.999mm)
进给功能	F	表示进给速度(1~1000mm/min)
主轴功能	S	表示主轴转速(0~9999r/min)
刀具功能	T	表示刀具号(0~99)
辅助功能	M	冷却液开、关控制等辅助功能(0~99)
偏移号	H	表示偏移代号(0~99)

(续)

功能	文字码	含义
暂停	P、X	表示暂停时间(0～99999.999s)
子程序号及子程序调用次数	P	子程序的标定及子程序重复调用次数设定(1～9999)
宏程序变量	P、Q、R	变量代号

表Ⅲ.2 M代码

M代码	功 能	M代码	功 能
M00	程序停止(自动循环)	M01	条件程序停止
M02	程序结束	M03	主轴正转
M04	主轴反转	M05	主轴停止
M06	刀具交换		
M08	冷却开	M09	冷却关
M18	主轴定向解除	M19	主轴定向
M29	刚性攻螺纹	M30	程序结束并返回程序头
M98	调用子程序	M99	子程序结束返回/重复执行

Ⅲ.2 FANUC数控车削编程

以FANUC 0i MATE-TB系统为例说明车削编程常用指令(表Ⅲ.3)。

表Ⅲ.3 FANUC 0i MATE-TB系统车削G代码

G代码	功 能	G代码	功 能
G00	快速定位(快速进给)	G01	直线插补(切削进给)
G02	顺时针(CW)圆弧插补	G03	逆时针(CCW)圆弧插补
G04	暂停(延时)	G10	可编程数据输入(资料设定)
G20	英制输入	G21	公制输入
G22	行程检查功能打开(ON)	G23	行程检查功能关闭(OFF)
G25	主轴速度波动检测断	G26	主轴速度波动检测开
G27	机械原点复位检查	G28	机械原点复位
G30	第二原点复位	G32	等螺距螺纹切削
G33	变螺距螺纹切削	G40	取消刀尖半径补偿
G41	刀尖半径左补偿	G42	刀尖半径右补偿

（续）

G 代码	功　　能	G 代码	功　　能
G50	工件坐标系、主轴最大速度设定	G52	局部坐标系设定
G53	机床坐标系设定		
G54	工件坐标系选择 1	G55	工件坐标系选择 2
G56	工件坐标系选择 3	G57	工件坐标系选择 4
G58	工件坐标系选择 5	G59	工件坐标系选择 6
G65	调用宏程序	G66	模态宏程序调用
G67	模态宏程序调用取消		
G70	精车固定循环	G71	外径粗车循环
G72	端面粗车循环	G73	固定形状粗车循环
G74	端面沟槽复合循环或深孔钻循环	G75	外径断续切槽循环
G76	多头螺纹切削循环	G90	外圆切削循环
G92	螺纹切削循环	G94	端面切削循环
G96	恒线速度控制有效	G97	恒线速度控制取消
G98	进给速度：每分进给（mm/min）	G99	进给速度：每转进给（mm/r）

1. G01 直线插补指令

格式：G01 X(U)__ Z(W)__ F__

1）圆角自动过渡

格式：G01 X__ R__ F__
　　　G01 Z__ R__ F__

说明：X 轴向 Z 轴过渡倒圆（凸弧）R 值为负，Z 轴向 X 轴过渡倒圆（凹弧）R 值为正。

例（图Ⅲ.1）：

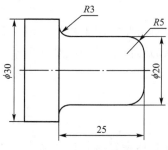

图Ⅲ.1　圆角自动过渡

O4001
N10 T0101;
N20 G0 X0 Z1. S900 M03;
N30 G1Z0. F0.2;

N40 G1 X20. R-5.;
N50 G1 Z-25. R3.;
N60 G1 X30.5;
N70 G28 X120. Z100.;
N80 M30;

2）直角自动过渡

格式：G01 X__ C__ F__
　　　G01 Z__ C__ F__

说明：倒直角用指令C，其符号设置规则同倒圆角。

例（图Ⅲ.2）：

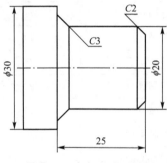

图Ⅲ.2　直角自动过渡

O4002
N10 T0101;
N20 G0 X0. Z1. S900 M03;
N30 G1Z0. F0.2;
N40 G1 X20. C-2.;
N50 G1 Z-25. R3.;
N60 G1 X30.5;
N70 G28 X120. Z100.;
N80 M30;

提示：自动过渡倒直角和圆角指令在用于精加工编程时会带来方便，但要注意符号的正负要准确，否则会发生不正确的动作。另外，某些FANUC系统倒直角采用I和K指令来表示C值。

2. G04暂停指令

格式：G04 X(U)__ 或 G04 P__

说明：指令中出现X、U或P均指延时，X和U用法相同，在其后跟延时时间，单位是秒，其后需加小数点。P后面的数字为整数，单位是ms。如需延时2s，该指令可表述为：G04 X2.0 或 G04 U2.0 或 G04 P2000。该指令尤其适于切槽时槽底加工暂停。

3. G28返回参考位置指令

格式：G28 X(U)__ Z(W)__

说明：G28指令意义类似于西门子G74指定。它的作用效果是各轴以快速移动速度通过中间点回参考点。它与G74指令不同的是G28指令中的坐标字有效，此位置作为中间点。指定语句 G28 U0 W0 即直接回参考点（中间点为程序执行前坐标）。

4. G32 等螺距螺纹加工指令

格式：G32 X(U)__ Z(W)__ F__

说明：G32指定为单刀切削螺纹指令，其中X、Z坐标指螺纹终点坐标。F指螺距，对端面螺纹，螺距采用半径值。

提示：G32指令也可用于加工连续螺纹切削。

5. G34 变螺距螺纹加工指令

格式：G34 X(U)__ Z(W)__ F__ K__

说明：G34螺纹用于加工增螺距螺纹或减螺距螺纹。所谓变螺距螺纹指的是以螺纹切入开始指定基准螺距值F，然后每隔一个螺距产生一个螺距差值（增值或减值）。

6. G50 坐标系设定或主轴最大速度设定指令

说明：G50指定用于在程序中设定编程坐标系原点的位置，即预置寄存指令。大多数系统使用G92指令作为预置寄存指令。G50指令格式和使用方法与G92相同。

G50也可用在恒线速度加工限制主轴最高转速。指令格式为"G50 S__"。

7. G90 外圆切削循环

1）车削圆柱面

格式：G90 X(U)__ Z(W)__ F__

说明：本指令的意义是在刀具起点与指定的终点间形成一个封闭的矩形。刀具从起点按先X方向起刀走一个矩形循环。其中，第一步和最后一步为G00动作方式，中间两步为G01动作方式，指令中的F字只对中间两步起作用。按刀具走刀方向，第一刀为G00方式动作；第二刀切削工件外圆；第三刀切削工件端面；第四刀G00方式快速退刀回起点。

例（图Ⅲ.3）：

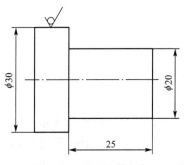

图Ⅲ.3　G90 外圆切削

O4003
N10 T0101;
N20 G0 X31. Z1. S800 M03;　　　（快速走刀至循环起点）

N30 G90 X26. Z-24.9 F0.3;   （X方向切深单边量2mm，端面留余量0.1mm精加工）
N40 X22.;   （G90模态，X向切深至22mm）
N50 X20.5;   （X向单边余量0.25mm精加工）
N70 X20. Z-25. F0.2 S1200;   （精车）
N80 G28 X100. Z100.;
N90 M30;

提示：因G90动作的第一刀为快速走刀，应注意起点的位置以确认安全。

2）车削圆锥面

格式：G90 X(U)__ Z(W)__ R__ F__

说明：R字代表被加工锥面的大小端直径差的1/2，即表示单边量锥度差值。对外径车削，锥度左大右小，R值为负，反之为正。对内孔车削，锥度左小右大，R值为正，反之为负。

例（图Ⅲ.4）：

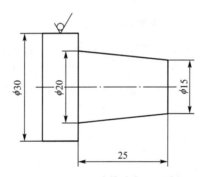

图Ⅲ.4　G90外锥度加工示例

O4004
N10 T0101;
N20 G0 X32. Z0.5 S900 M3;   （刀具定位）
N30 G90 X26. Z-25. R-2.5 F0.15;   （粗加工）
N40 X22.;
N50 X20.5;   （留精加工余量双边0.5mm）
N60 G0 Z0 S1200 M3;
N70 G90 X20. Z-25. R-2.5 F0.1;
N80 G28 X100. Z100.;
N90 M5;
N100 M2;

注意：在进行锥面精加工时，注意刀具起始位置的Z轴坐标应与实际锥度的起点Z坐标一致，否则加工出的锥度不正确；若刀具起始位置的Z轴坐标取值与实际锥度的起点Z坐标不一致，则应算出锥面轮廓延长线上对应所取Z坐标处与锥面终点处的实际直径差。

8．G92螺纹切削循环指令

1）圆柱螺纹加工

格式：G92 X(U)__ Z(W)__ F__

说明：加工过程中，刀具先沿 X 轴进刀至 X(U) 坐标；第二步沿 Z 轴切削螺纹，当到达某一位置时，接收到从机床发来的信号，起动螺纹倒角，到达 Z(W) 坐标；第三步刀具沿 X 轴退刀至 X 初始坐标；第四步沿 Z 轴退刀至 Z 初始坐标，加工结束。螺纹倒角距离在 0.1L 至 12.7L 之间指定（L 为螺距），指定单位为 0.1L（两位数：从 00 到 99），由参数 5130 决定。

例（图Ⅲ.5）：

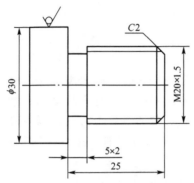

图Ⅲ.5　G92 外圆柱螺纹加工

```
O4005
N110 T0303;                （仅螺纹加工段）
N120 G0 X28. Z5. S350 M3;  （刀具定位）
N130 G92 X19.4 Z-23. F1.5; （螺纹加工）
N140 X19.;                 （逐层进刀）
N150 X18.6;
N160 X18.2;
N170 X18.;
N180 X17.9;
N190 X17.8;
…
```

2）车削圆锥螺纹

格式：G92 X(U)__ Z(W)__ R__ F__

说明：R 字代表被加工锥螺纹的大小端外径差的 1/2，即表示单边量锥度差值。对外螺纹车削，锥度左大右小，R 值为负，反之为正；对内螺纹车削，锥度左小右大，R 值为正，反之为负。加工过程与圆柱螺纹的加工相同。

例（图Ⅲ.6，$P=1.5$mm）：

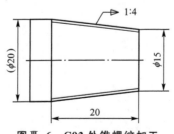

图Ⅲ.6　G92 外锥螺纹加工

O4006
N10 T0101;
N20 G0 X25.Z5.S900 M3;
N30 G92 X19.6 Z-20.R-2.5 F1.5;
N40 X19.4;
N50 X19.;

9. G94 端面切削循环指令

1) 端面加工

格式：G94 X(U)__Z(W)__F__

说明：本指令主要用于加工长径比较小的盘类工件，它的车削特点是利用刀具的端面切削刃作为主切削刃。G94 区别于 G90，它是先沿 Z 方向快速走刀，再车削工件端面，退刀光整外圆，再快速退刀回起点。按刀具走刀方向，第一刀为 G00 方式动作快速进刀；第二刀切削工件端面；第三刀 Z 方向退刀切削工件外圆；第四刀 G00 方式快速退刀回起点。

例（图Ⅲ.7）：

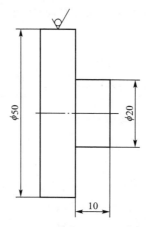

图Ⅲ.7 G94 端面加工示例

O4007
N10 T0101;
N20 G0 X52.Z1.S900 M03;
N30 G94 X20.2 Z-2.F0.2;   （粗车第一刀，Z 向切深 2mm）
N40 Z-4.;
N50 Z-6.;
N60 Z-8.;
N70 Z-9.8;
N80 X20.Z-10.S1200;   （精加工）
N90 G28 X100.Z100.;
N100 M30;

2) 锥面加工

格式：G94 X(U)__Z(W)__R__F__

说明：注意 G94 和 G90 加工锥度轴意义有所区别，G94 是在工件的端面上形成斜面，而 G90 是在工件的外圆上形成锥度。

指令中 R 字表示为圆台的高度。圆台左大右小，R 为正值；若则圆台直径左小右大，则 R 为负值，一般只在内孔中出现此结构，但用镗刀 X 向进刀车削并不妥当。

提示：上述 G90、G94 二指令中的 X、Z 字均指与起刀点相对的对角点的坐标。

10. 多重复合循环

FANUC 系统提供多重复合固定循环指令，主要用于粗、精车外形、内孔、钻孔、切槽、螺纹等加工，可以大大简化编程。G71、G72 和 G73 主要用于毛坯的粗车；G70 用于精车；G74 和 G75 用于切槽和钻孔；G76 用于螺纹加工循环。

1) G70 精车固定循环指令

格式：G70 P(ns)＿Q(nf)＿

说明：G70 指令用于在 G71、G72、G73 指令粗车工件后进行精车循环。在 G70 状态下，在指定的精车描述程序段中的 F、S、T 有效。若不指定，则维持粗车前指定的 F、S、T 状态。G70 到 G73 中 ns 到 nf 间的程序段不能调用子程序。当 G70 循环结束时，刀具返回到起点并读下一个程序段。

关于 G70 的详细应用请参见 G71、G72 和 G73 部分。

2) G71 外径粗车循环指令

格式：G71 U(Δd)＿R(e)＿
G71 P(ns)＿Q(nf)＿U(Δu)＿W(Δw)＿F＿S＿T＿

Δd——循环每次的切削深度（半径值、正值）；

e——每次切削退刀量；

ns——精加工描述程序时开始循环程序段的行号；

nf——精加工描述程序时结束循环程序段的行号；

u——X 向精车预留量；

w——Z 向精车预留量。

说明：G71 指令称之为外径粗车固定循环，它适用于毛坯料粗车外径和粗车内径。在 G71 指令后描述零件的精加工轮廓，CNC 系统根据加工程序所描述的轮廓形状和 G71 指令内的各个参数自动生成加工路径，将粗加工待切除余量一次性切削完成。

例（图Ⅲ.8）：

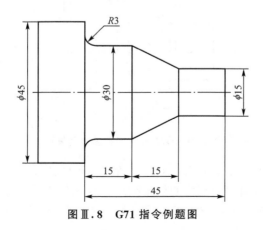

图Ⅲ.8　G71 指令例题图

O4008
N10 T0101;
N20 G0 X46. Z0.5 S500 M03;
N30 G71 U2. R0.5;   （每层切深 2mm，退刀 0.5mm）
N40 G71 P50 Q110 U0.3 W0.1 F0.3; （精加工余量 X 向单边量 0.3mm，Z 向 0.1mm。粗切进给量 0.3mm/r）
N50 G1 X15.;
N60 G1 Z0. F0.15 S800;   （精加工进给量 0.15mm/r。精车转速为 800r/min）
N70 Z-15.;
N80 X30. Z-30.;
N90 Z-42.;
N100 G2 X36. Z-45. R3.;
N110 G1 X46.;
N120 G70 P50 Q100;   （精加工循环）
N130 G28 X100. Z100.;
N140 M5;
N150 M30;

3）G72 端面粗车循环指令

格式：G72 W(d)__R(e)__
　　　G72 P(ns)__Q(nf)__U(u)__W(w)__F__S__T__

d——循环每次的切削深度（正值）；

e——每次切削退刀量；

ns——精加工描述程序时开始循环程序段的行号；

nf——精加工描述程序时结束循环程序段的行号；

u——X 向精车预留量；

w——Z 向精车预留量。

G72 指令内部参数示意图如图Ⅲ.9 所示。

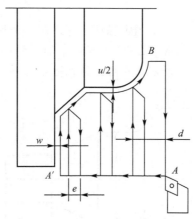

图Ⅲ.9　G72 指令内部参数示意图

说明：端面粗车循环指令的含义与 G71 类似，不同之处是刀具平行于 X 轴方向切削，它是从外径方向往轴心方向切削端面的粗车循环，该循环方式适于对长径比较小的盘类工件端面方向粗车。和 G94 一样，对 93°外圆车刀，其端面切削刃为主切削刃。

注意：
（1）G72不能用于加工端面内凹的形体；
（2）精加工首刀进刀须有Z向动作；
（3）循环起点的选择应在接近工件处以缩短刀具行程和避免空走刀。
例（图Ⅲ.10）：

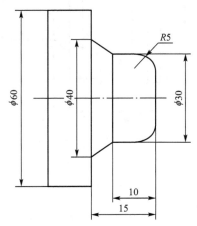

图Ⅲ.10　G72指令例题图

O4009
N10 T0101;
N20 G0 X61.Z0.5 S500 M03;
N30 G72 W2.R0.5;
N40 G72 P50 Q100 U0.1 W0.3 F0.25;
N50 G0 Z-15.;
N60 G1 X40.F0.15 S800;
N70 X30.Z-10.;
N80 Z-5.;
N90 G2 X20.Z0 R5.;
N100 G0 Z0.5;
N110 G70 P60 Q110;
N120 G28 X100.Z100.;
N130 M30;

4）固定形状粗车循环指令

格式：G73 U(Δi)__ W(Δk)__ R(Δd)__
　　　　G73 P(ns)__ Q(nf)__ U(Δu)__ W(Δw)__ F__ S__ T__

Δi——X方向毛坯切除余量（半径值、正值）；

Δk——Z方向毛坯切除余量（正值）；

Δd——粗切循环的次数；

ns——精加工描述程序时开始循环程序段的行号；

nf——精加工描述程序时结束循环程序段的行号；

Δu——X向精车预留量；

Δw——Z 向精车预留量。

例：加工如图Ⅲ.11所示的工件，其毛坯为锻件。工件 X 向残留余量不大于 5mm。Z 向残留余量不大于 3mm。要求采用 G73 方式切削出该零件。

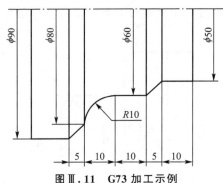

图Ⅲ.11　G73 加工示例

```
O4010
N10 T0101;
N20 G0 X110. Z10. S800 M3;
N30 G73 U5. W3. R3. ;
N40 G73 P50 Q110 U0. 4 W0. 1 F0. 3;
N50 G0 X50. Z1. S1000;
N60 G1 Z-10. F0. 15;
N70 X60. Z-15. ;
N80 Z-25. ;
N90 G2 X80. Z-35. R10. ;
N100 G1 X90. Z-40. ;
N110 G0 X110. Z10. ;
N120 G70 P50 Q110;
N130 G28 X100. Z150. ;
N140 M30;
```

5）G74 端面沟槽复合循环或深孔钻循环指令

格式：G74 R(e)＿

　　　G74 X(u)＿Z(w)＿P(Δi)＿Q(Δk)＿R(Δd)＿F＿

e——每次退刀量；

u——X 向终点坐标值；

w——Z 向终点坐标值；

Δi——X 向每次的移动量，P1000 为 1mm；

Δk——Z 向每次的切入量；

Δd——切削到终点时的 X 轴退刀量（可以缺省）；

F——进给量。

注：X 向终点坐标值为实际 X 向终点尺寸减去双边刀宽。

说明：该指令可实现端面深孔和端面槽的断屑加工，Z 向切进一定的深度，再反向退刀一定的距离，实现断屑。指定 X 轴地址和 X 轴向移动量，就能实现端面槽加工；若不指定 X 轴地址和 X 轴向移动量，则为端面深孔钻加工。

例：端面切槽（图Ⅲ.12）

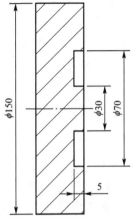

图Ⅲ.12　G74指令例题图

O4011
N10 T0606;
N20 S900 M3;
N30 G0 X30. Z2.;
N40 G74 R1.;
N50 G74 X62. Z-5. P3500 Q3000 F0.1;
N60 G0 X200. Z50. M5;
N70 M30;

（端面切槽刀，刃口宽 4mm）

6）G75 外径断续切槽循环指令

格式：G75 R(e)__
　　　G75 X(u)__Z(w)__P(Δi)__Q(Δk)__R(Δd)__F__

e——分层切削每次退刀量；
u——X 向终点坐标值；
w——Z 向终点坐标值；
Δi——Z 向每次的切入量；
Δk——X 向每次的移动量；
Δd——切削到终点时的退刀量(可以缺省)。

例：切削较宽的径向槽（图Ⅲ.13）

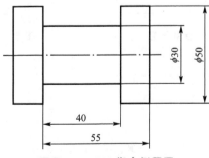

图Ⅲ.13　G75指令例题图

O4012
N10 T0202;   （切槽刀，刃口宽 5mm)
N20 S800 M3;
N30 G0 X52. Z-15.;
N40 G75 R1.;
N50 G75 X30. Z-50. P3000 Q4500 F0.1;
N60 G0 X150. Z100. M5;
N70 M30;

7）G76 多头螺纹切削循环指令

格式：G76 P(m)(r)(a)__Q(Δdmin)__R(d)__
　　　G76 X(U)__Z(W)__R(i)__P(k)__Q(Δd)__F(L)__

m——精加工重复次数（1~99）。该值是模态的，此值可以用 5142 号参数设定，由程序指令改变。

r——倒角量。当螺距由 L 表示时，可以从 0.0L 到 9.9L 设定，单位为 0.1L（两位数：从 00 到 99）。该值是模态的，此值可用 5130 号参数设定，由程序指令改变。

a——刀尖角度。可以选择 80°、60°、55°、30°、29°和 0°六种中的一种，由 2 位数规定。该值是模态的。可用参数 5143 号设定，用程序指令改变。

d——精加工余量。该值是模态的，可用 5141 号参数设定，用程序指令改变。

i——螺纹半径差。如果 i=0，可以进行普通直螺纹切削。

k——螺纹高。此值用半径规定。

Δd——第一刀切削深度（半径值）。

L——螺距（同 G32）。

G76 指令参数示意图如图Ⅲ.14 所示。

例：G76 指令外螺纹加工（图Ⅲ.15）。

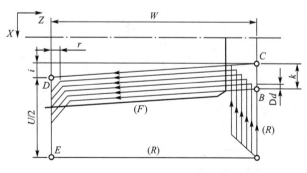

图Ⅲ.14　G76 指令参数示意图

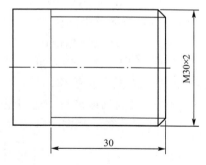

图Ⅲ.15　G76 指令螺纹加工例题图

O4013
N10 T0303;
N20 S800 M3;
N30 G0 X35. Z3.;
N40 G76 P021260 Q100 R100;   （螺纹参数设定，R 为正）
N50 G76 X26.97 Z-30. R0 P1510 Q200 F2.;

N60 G0 X100.Z100.M5;
N70 M2;

例：G76指令内螺纹加工编程示例（图Ⅲ.16）

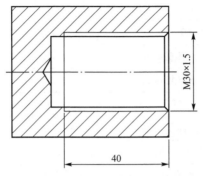

图Ⅲ.16　G76指令内螺纹加工例题图

O4014
N10 T0303;
N20 S800 M3;
N30 G0 X25.Z4.;
N40 G76 P021060 Q100 R-100;（螺纹参数设定，R为负）
N50 G76 X30.Z-40.P9742 Q200 F1.5;
N60 G0 X100.Z100.;
N70 M5;
N80 M2;

Ⅲ.3　FANUC 数控铣削编程

以 FANUC 0MC 数控系统为例。该系统扩展后可联动控制轴数为四轴；编程代码通用性强，编程方便，可靠性高。机床可供用户使用的 G 代码列表见表Ⅲ.4。

表Ⅲ.4　FANUC 0MC 数控系统铣削 G 代码

G 代码	功　　能	G 代码	功　　能
G00	快速定位（快速进给）	G01	直线插补（切削进给）
G02	顺时针（CW）圆弧插补	G03	逆时针（CCW）圆弧插补
G04	暂停（延时）	G09	精确停止
G10	资料设定	G11	资料设定模式取消
G15	极坐标指令取消	G16	极坐标指令
G17	XY 平面选择	G18	ZX 平面选择
G19	YZ 平面选择	G20	英制输入

（续）

G代码	功　　能	G代码	功　　能
G21	公制输入	G22	行程检查功能打开(ON)
G23	行程检查功能关闭(OFF)	G27	机械原点复位检查
G24	镜像功能	G25	取消镜像功能
G28	机械原点复位	G29	从参考原点自动复位
G30	第二原点复位	G31	跳跃功能
G33	螺纹切削	G39	转角补正圆弧切削
G40	刀具半径补偿取消	G41	刀具半径左补偿
G42	刀具半径右补偿	G43	刀具长度正补偿
G44	刀具长度负补偿	G49	刀具长度补偿取消
G50	取消比例缩放	G51	比例缩放
G52	局部坐标系设定	G53	机床坐标系选择
G54	工件坐标系选择1	G55	工件坐标系选择2
G56	工件坐标系选择3	G57	工件坐标系选择4
G58	工件坐标系选择5	G59	工件坐标系选择6
G65	调用宏程序	G66	模态宏程序调用
G67	模态宏程序调用取消	G68	坐标系旋转
G69	取消坐标系旋转	G73	高速深孔钻孔循环
G74	左旋螺纹攻螺纹循环	G76	精镗孔循环
G80	固定循环取消	G81	钻孔循环(钻中心孔)、钻镗孔
G82	钻孔循环(带停顿)、反镗孔	G83	深孔钻孔循环
G84	攻螺纹循环	G85	粗镗孔循环
G86	镗孔循环	G87	反镗孔循环
G90	绝对坐标	G91	增量坐标
G92	加工坐标系设定	G98	固定循环返回起始点

1. G90绝对坐标指令和G91增量坐标指令

G90绝对坐标方式：在该方式下，程序段中的尺寸为绝对坐标值。

G91增量坐标方式：在该方式下，程序段中的尺寸为增量坐标值，即相对于前一工作点的增量值。

2. G00快速进刀指令

格式：G00 X__ Y__ Z__

说明：最快进给速度为系统默认，由系统参数调整。

3. G01 直线插补指令

格式：G01 X__ Y__ Z__ F__

说明：G90 和 G91 时，起刀的原点不同，F 单位是 mm/min。

4. G02/G03 圆弧插补指令

格式：G02 X__ Y__ R__ F__
　　　G02 X__ Y__ I__ J__ F__

说明：在 X__ Y__ 平面(G17)内顺时针圆弧插补，若在 ZX 平面(G18)或 YZ 平面(G19)改变相应坐标字。

采用 G90 时，X__ Y__ Z__ 是圆弧终点相对于工件零点的坐标。采用 G91 时，X__ Y__ Z__ 是圆弧终点相对于圆弧起点的坐标。I__ J__ K__ 是圆弧的圆心坐标值，均为圆心点相对于圆弧起点的增量值。R__ 为圆弧半径。圆心角小于等于 180°R 为正值、圆心角大于 180°R 为负值。当圆弧为整圆时，不能用 R__，只能用 I__ J__ K__。

5. G04 进给暂停指令

格式：G04 X__ 或 G04 P__

说明：G04 指令可使进给暂停，刀具在某一点停留一段时间后再执行下一段程序。X 或 P 均为指定进给暂停时间。两者区别是：X 后面可带小数点，单位是 s；P 后面数字不能带小数点，单位是 ms。如，G04 X3.5，或者 G04 P3500，都表示刀具暂停了 3.5s。

6. 刀具半径补偿指令

格式：G41(G42)G01 X__ Y__ D__

D——刀具号，存有预先由 MDI 方式输入的刀具半径补偿值。
G41 为左刀补指令，表示沿着刀具进给方向看，刀具中心在零件轮廓的左侧；
G42 为右刀补指令，表示沿着刀具进给方向看，刀具中心在零件轮廓的右侧。
G40 为取消刀具半径补偿指令。

格式：G40 G01 X__ Y__

说明：G40 与 G41 或 G42 要成对使用；从无刀补状态进入刀补状态转换时必须采用 G00 或 G01 直线移动指令，不能用 G02、G03；刀补撤销时也要用 G00 或 G01 直线移动指令。

7. G43、G44、G49 刀具长度补偿指令

格式：G43(G44)G01 Z__ H__

　　　　　H——刀具号，存有预先由 MDI 方式输入的刀具长度补偿值。
G43 为正补偿，表示刀具在 Z 方向实际坐标值比程序给定值增加一个偏移量；
G42 为负补偿，表示刀具在 Z 方向实际坐标值比程序给定值减少一个偏移量。

格式：G49 G01 Z__

说明：G49 为取消刀具长度补偿指令。

8. G92 设置加工坐标系指令

格式：G92 X__ Y__ Z__

说明：将加工原点设定在相对于刀具起始点的某一空间点上。

例：G92 X20.Y10.Z10.；

其确立的加工原点在距离刀具起始点 $X=-20$，$Y=-10$，$Z=-10$ 的位置上，如图Ⅲ.17 所示。

9. G53 选择机床坐标系指令

格式：G53 G90 X__Y__Z__

G53 指令使刀具快速定位到机床坐标系中的指定位置上，式中 X、Y、Z 后的值为机床坐标系中的坐标值，其尺寸均为负值。

例：G53 G90 X-100.Y-100.Z-20.；

执行后刀具在机床坐标系中的位置如图Ⅲ.18 所示。

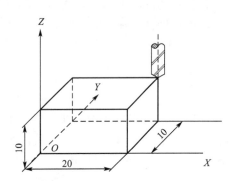
图Ⅲ.17　G92 设置加工坐标系

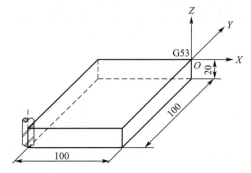
图Ⅲ.18　G53 选择机床坐标系

10. G54、G55、G56、G57、G58、G59 选择 1～6 号加工坐标系指令

格式：G54 G90 G00（G01）X__Y__Z__（F__）

说明：这些指令可以分别用来选择相应的加工坐标系，该指令执行后，所有坐标值指定的坐标尺寸都是选定的工件加工坐标系中的位置。1～6 号工件加工坐标系是通过 CRT/MDI 方式设置的。

例：在图Ⅲ.19 中，用 CRT/MDI 在参数设置方式下设置了两个加工坐标系：

G54：X-50.Y-50.Z-10.；
G55：X-100.Y-100.Z-20.；

这时，建立了原点在 O' 的 G54 加工坐标系和原点在 O'' 的 G55 加工坐标系。若执行下述程序段：

N10 G53 G90　　X0.Y0.Z0.；
N20 G54 G90 G01 X50.Y0.Z0.F100；
N30 G55 G90 G01 X100.Y0.Z0.F100；

则刀尖点的运动轨迹如图Ⅲ.19 中 OAB 所示。

例：使用半径为 5mm 的刀具加工如图Ⅲ.20 所示的零件，加工深度为 5mm，加工程序编制如下：

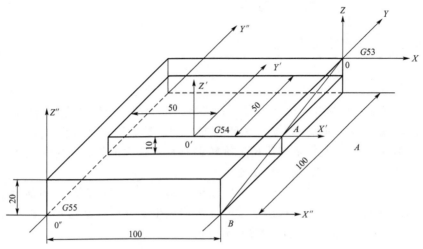

图Ⅲ.19 设置加工坐标系

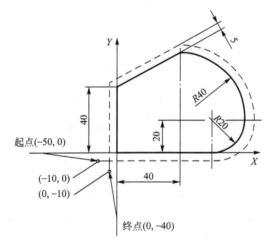

图Ⅲ.20 零件图样

设置 G55：X=-400.，Y=-150.，Z=-50.；H01=5。
O1001
G55 G90 G01 Z40.F200;   （2号加工坐标系）
M03 S500;   （主轴启动）
G01 X-50.Y0.;   （到达 X、Y 坐标起始点）
G01 Z-5.F100;   （到达 Z 坐标起始点）
G01 G42 X-10.Y0.H01;   （建立右偏刀具半径补偿）
G01 X60.Y0.;   （切入轮廓）
G03 X80.Y20.R20.;   （切削轮廓）
G03 X40.Y60.R40.;   （切削轮廓）
G01 X0.Y40.;   （切削轮廓）
G01 X0.Y-10.;   （切出轮廓）
G01 G40 X0.Y-40.;   （撤销刀具半径补偿）
G01 Z40.F200;   （Z 坐标退刀）
M05;   （主轴停）

M30;   （程序停）

11. G68、G69 坐标系旋转功能

格式：G68 X__ Y__ R__
      G69

X、Y——旋转中心的坐标值。当 X、Y 省略时，当前的位置即为旋转中心；

R——旋转角度，逆时针旋转定义为正方向，顺时针旋转定义为负方向。

说明：该指令可使编程图形按照指定旋转中心及旋转方向旋转一定的角度，G68 表示开始坐标系旋转，G69 用于撤销旋转功能。

当程序在绝对坐标方式下时，G68 程序段后的第一个程序段必须使用绝对坐标方式移动指令，才能确定旋转中心。如果这一程序段为增量坐标方式移动指令，那么系统将以当前位置为旋转中心，按 G68 给定的角度旋转坐标。

例（图Ⅲ.21）：

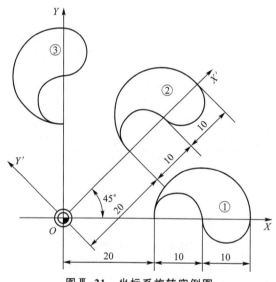

图Ⅲ.21　坐标系旋转实例图

O1002   （主程序）
N10 G90 G17 S800 M03;
N20 M98 P1100;   （加工①）
N30 G68 X0.Y0.P45;   （旋转 45°）
N40 M98 P1100;   （加工②）
N50 G69;   （取消旋转）
N60 G68 X0.Y0.P90;   （旋转则 90°）
M70 M98 P1100;   （加工③）
N80 G69 M05 M30;   （取消旋转）
子程序(①的加工程序)
O1100
N100 G90 G01 X20.Y0.F100;
N110 G02 X30.Y0.R5.;

N120 G03 X40.Y0.R5.;
N130 X20.Y0.R10.;
N140 G00 X0.Y0.;
N150 M99;

12. M98 子程序调用指令

格式：M98 P__

P——表示子程序调用情况。P后共有8位数字，前四位为调用次数，省略时为调用一次；后四位为所调用的子程序号。

说明：编程时，为了简化程序的编制，当一个工件上有相同的加工内容时，常用调用子程序的方法进行编程。调用子程序的程序叫作主程序。子程序的编号与一般程序基本相同，只是程序结束字为M99表示子程序结束，并返回到调用子程序的主程序中。

13. G51 比例功能指令

(1) 比例功能

格式：G51 X__Y__Z__P__
　　　G50

X、Y、Z——比例中心坐标(绝对坐标方式)；

P——比例系数，最小输入量为0.001，比例系数的范围为：0.001～999.999。该指令以后的移动指令，从比例中心点开始，实际移动量为原数值的P倍。P值对偏移量无影响。

说明：比例及镜向功能可使原编程尺寸按指定比例缩小或放大；也可让图形按指定规律产生镜像变换。G51为比例编程指令；G50为撤销比例编程指令。

(2) 各轴以不同比例编程

格式：G51 X__Y__Z__I__J～K__
　　　G50

X、Y、Z——比例中心坐标；

I、J、K——对应X、Y、Z轴的比例系数，在±0.001～±9.999范围内。本系统设定I、J、K不能带小数点，比例为1时，应输入1000，并在程序中都应输入，不能省略。比例系数与图形的关系如图Ⅲ.22所示。其中：b/a：X轴系数；d/c：Y轴系数；O：比例中心。

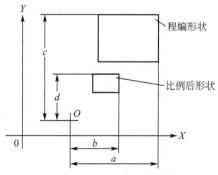

图Ⅲ.22　各轴以不同比例编程

14. G24/G25 镜像功能指令

各个轴镜像加工功能。G25 为取消镜像。

例（图Ⅲ.23）：

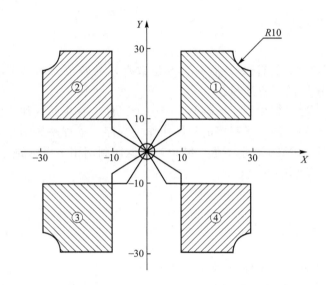

图Ⅲ.23　镜像加工实例图

```
O1003                    （主程序）
N10 G91 G17 S800 M03;
N20 M98 P1101;           （加工①）
N30 G24 X0.;             （Y轴镜像，镜像位置为X=0）
N40 M98 P1101;           （加工②）
N50 G24 X0.Y0.;          （X轴、Y轴镜像，镜像位置为(0，0)）
N60 M98 P1101;           （加工③）
N70 G25 X0.;             （取消Y轴镜像）
N80 G24 Y0.;             （X轴镜像）
N90 M98 P1101;           （加工④）
N100 G25 Y0.;            （取消镜像）
N110 M05;
N120 M30;
```

子程序（①的加工程序）：

```
%1101
N200 G41 G00 X10.0 Y4.0 D01;
N210 Y1.0
N220 Z-98.0;
N230 G01 Z-7.0 F100;
N240 Y25.0;
N250 X10.0;
N260 G03 X10.0 Y-10.0 I10.0;
```

N270 G01 Y-10.0;
N280 X-25.0;
N290 G00 Z105.0;
N300 G40 X-5.0 Y-10.0;
N310 M99;

References

[1] James V Valentino, Joseph Goldenberg. Introduction to Computer Numerical Control [M]. 5th ed. Upper Saddle River: Prentice Hall, 2012.

[2] Posinasetti Nageswara Rao. CAD/CAM: Principles and Applications [M]. 3rd ed. Bombay: Tata McGraw-Hill Education Pvt. Ltd. , 2010.

[3] The Hong Kong Polytechnic University. Computer Numerical Control (CNC) [EB/OL]. [2009-12-20]. http://www2.ic.polyu.edu.hk/student _ net/training _ materials/IC% 20Workshop% 20Materials%2009%20-%20Computer%20Numerical%20Control%20 (CNC).pdf

[4] Pang Tze Hong. Practical Development of an Open Architecture Personal Computer-based Numerical Control (OAPC-NC) System [D]. Hattiesbury: The University of Sourthern Mississippi, 2005.

[5] Hu Zhan-qi. Numerical Control Technology [M]. Wuhan: Wuhan University of Technology Press, 2004.

[6] Mahbubur Rahman. Modeling and Measurement of Multi-axis Machine Tools to Improve Positioning Accuracy in a Software Way [D]. Oulu: Oulu University, 2004.

[7] Alan Overby. CNC Machining Handbook: Building, Programming, and Implementation [M]. New York: McGraw-Hill/TAB Electronics, 2010.

[8] Suk-Hwan Suh, Seong Kyoon Kang, Dae-Hyuk Chung, Ian Stroud. Theory and Design of CNC System [M]. Berlin: Springer, 2010.

[9] Viera Poppeova, Vladimir Bulej, Juraj Uricek, Monika Rupiková. The Development of Hexapod Kinematic Machine [J]. Journal of Trends in the Development of Machinery and Associated Technology, 2012, 16(1), 39-42.

[10] Peter Smid. CNC Control Setup for Milling and Turning: Mastering CNC Control Systems [M]. South Norwalk: Industrial Press, Inc. , 2010.

[11] 吴瑞明. 数控技术 [M]. 北京：北京大学出版社, 2012.